Solved Problems in Well Testing

Iraj Ershaghi

Solved Problems in Well Testing

Quantitative Geology

Springer

Iraj Ershaghi
University of Southern California
Los Angeles, CA, USA

ISBN 978-3-031-47298-5 ISBN 978-3-031-47299-2 (eBook)
https://doi.org/10.1007/978-3-031-47299-2

This Springer imprint is published by the registered company Springer Nature Switzerland AG
The registered company address is: Gewerbestrasse 11, 6330 Cham, Switzerland

Paper in this product is recyclable.

To my wife Mitra for her unconditional support, and to my children: Marsha, Minta, and Milad.

Preface

This book is about the technology of measuring pressure responses by changing the production or injection rate from wellbores producing oil, gas, geothermal fluids, and water. Relating production or injection rate changes to reservoir pressure responses allows the estimation of subsurface reservoir properties and mapping of the subsurface reservoir rock and fluid interfaces. This book aims at orienting the subsurface geologist and industry professionals to the understanding the well testing as a supporting tool in reservoir characterization and in the realization of the power of using well testing in enhancing subsurface characterization and modeling by including results from pressure transient tests. It is also intended for subsurface fluid storage professionals who need to quickly grasp the concepts and realize the information opportunity that the pressure transient test can offer in leak detection and inventory management.

The main credit for the technology of well-test interpretation goes to the work done by early research scientists who first recognized this opportunity many decades ago. Over the years, contributions from universities, industry and service companies experts have helped to solidify the concepts and have established the science and the art of well-test analyses. Now, this technology has become a powerful tool in the hands of subsurface characterization engineers and geoscientists. With the increasing interest in subsurface storage of carbon dioxide and compressed air and hydrogen, monitoring of pressure data can serve an important purpose in the management of subsurface changes and detection of leaks.

At one time, the application of well-testing technology was limited to analyzing simple conventional wells in assumed homogeneous reservoirs. Then as more innovations were implemented in the drilling and completion design of boreholes, new formulations were developed to address such complex systems. Among the developments were the recognition of complexities associated with heterogeneous reservoirs and the realization of permeability trends caused by complex geological factors and conditions, incorporation of wellbore alterations by the skin and induced fractures, and the handling of boundary effects and reservoir discontinuities. Parallel to advancing interpretation methods, progress was also made in the design and manufacturing of measurement tools and techniques for obtaining more accurate data.

Such advances have made it possible now to extend the interpretation techniques to even more complex reservoir and wellbore conditions. Focus on horizontal drilling and later dealing with extremely tight reservoirs have resulted in new interpretation technologies. Simultaneous measurement of flow rate and pressure has opened new opportunities for a more precise look at subterranean reservoir characteristics. As production operations in many fields have switched to artificial lift methods, the technology has also been extended to focus on handling low-pressure reservoir cases. These solution methodologies have also provided new opportunities for regulatory agencies to monitor safe injection gradients for fluid injection, assisted recoveries, and fluid disposal purposes.

The art and science of well-test analysis include a qualitative pattern recognition of rate and pressure signals and a quantitative approach for estimating parameters. Advances in computer-aided methods, capabilities in data de-nosing, and the use of artificial intelligence (AI) and machine learning for pattern recognition have opened new opportunities in the technology of well testing. In recent years, in transient rate analysis, more and more emphasis has been focused on data-driven solutions for pattern recognition and prediction purposes.

Now it is widely recognized that well testing can be a powerful tool in the hands of subsurface reservoir analysts. These tests offer unique reservoir characterization capabilities. Such solution methodologies help in the estimation of key reservoir parameters that contribute to improved resource estimation and help in inverse modeling during reservoir simulation. These interpretation techniques also help in recognizing changes in phase behavior and rock properties caused by fluid movements and geomechanical effects around the wells and inside the reservoir.

I have been teaching graduate-level courses on the topic of well testing since 1972 and have also conducted and supervised research in this area. Over the years, I have made substantial changes to my teaching methodology to incorporate advances in the science and theory of pressure transient tests. Many of my students have been engineering students, but occasionally I have had some students with expertise in subsurface geology. Over the years, I have prepared and designed assignments to guide students and professionals in well-testing courses, helping them to better understand and appreciate the opportunities offered by relevant well-testing tools and interpretation systems. In every aspect of this technology, mathematics is involved, which can add a dimension of complexity. But one can develop an important appreciation of these methodologies without being drowned in associated mathematics. My goal in preparing this book is to offer an opportunity for new users to understand these concepts by reviewing the extensive set of solved problems in the Appendix of this book where I include a substantial number of problems and their brief solutions that can help the students and professionals understand various pressure transient testing concepts.

The book and the solved problems in the Appendix include a focus on key principles and can target a technologically evolving audience. The questions included point to specific references and specialized papers that define the focus areas. The offered solutions to the problems are described with the necessary graphs and illustrations. In some cases, a partial solution is included to allow the reader to further check the concepts described in this or other similar books and references. As my lecture materials in the past have included design cases I have developed using the Fekete IHS software and the Well Test by Kappa Engineering software, I acknowledge the significant educational value of their software products. It is my hope that this book, with its mission to uncomplicate the understanding of well testing, can help in the more widespread use of well-testing technology. These concepts help in evaluating new focus areas such as characterizing the pore spaces of subsurface geologic formations used as storage sites, designing, and prolonging the life of marginal wells, and monitoring subsurface integrity for greenhouse gas storage and, more importantly, energy storage to benefit the environmental safety purposes.

My parents get credit for teaching me the art of uncomplicating things. It is my hope that this book, with its mission to uncomplicate the understanding of well-test technology, can help in the more widespread use of well testing and help in prolonging the life of marginal wells and in monitoring reservoir integrity for subsurface storage.

Palos Verdes Estates, CA, USA Iraj Ershaghi, Ph.D., P.E.

Contents

Nomenclature and Legends

Throughout the book, certain symbols are used, the list of which is shown below. As for the units, the British system of units is used which is the norms in oil and gas fields in the USA.

Formation Petrophysics and Geomechanical Properties

k	Permeability, mD
φ	Porosity, fraction
h	Formation Thickness, ft
c_o	Oil Compressibility, 1/psi
c_w	Water Compressibility, 1/psi
c_g	Gas Compressibility, 1/psi
B_o	Formation Volume Factor, Rb/STB
B_w	Formation Volume Factor, Rb/STB
B_g	Formation Volume Factor, Rcf/SCF
R' or F'	t_{lag}/t_c
So	Oil Saturation, fraction
Sw	Water Saturation, fraction
Sg	Gas Saturation, fraction
μ	Dynamic Viscosity, cp

Test Parameters

q_o	Daily Rate of Oil Production, STB/D
q_w	Daily Rate of Water Production, STB/D
q_g	Daily Rate of Gas Production, SCF/D

Measured Parameters

P	Pressure, psi
T	Temperature, °F
Pwh	wellhead Pressure, psi
PBH	Bottomhole Pressure, °F

Estimated Parameters

C	Wellbore Storage, bbl/psi
s or sk	Skin Factor, dimensionless
S	Storativity, ft./psi
S	Storativity, (groundwater), dimensionless
T	Transmissibility, mD. ft./cp
T	Transmissivity, ground water, m^2/sec
t_C	Cycle Time in Pulse Testing

Additional Symbols for Naturally Fractured Reservoirs

λ	Interporosity Flow Coefficient
ω	Partition Storage Ratio

Specific Symbols for Hydraulically Fractured Wells

F	Fracture Conductivity, dimensionless
X_f	Fracture Extent, ft
W	Fracture Width, ft

Abbreviations

$(\frac{tp+dt}{dt})$	Horner Time
ANN	Artificial Neural Network
BHP	Bottomhole Pressure
CBM	Coalbed Methane
Dp or Delp	Pressure Change
DST	Drill Stem Test

dt or Delt	Time Increment
DTS	Differential Temperature Sensor
EFM	Electronic Flow Meter
EM	Electromagnetic
ESP	Electrical Submersible Pump
F	Fahrenheit
FHP	Final Hydrostatic Pressure
IHP	Initial Hydrostatic Pressure
LWD	Logging While Drilling
MDH	Miller–Dyes–Hutchinson
P.' dt	Pressure Derivative Function
PCP	Progressive Cavity Pump
PLT	Production Logging Tool
PNR	Pressure-Normalized Rate
PSS	Pseudo-steady State
PVT	Pressure–Volume–Temperature
RNP	Rate-Normalized Pressure
USS	Unsteady State

Chapter 1
Introduction

Charles Fairhurst (2017) in his article argues, “it is essential that we understand subsurface of the earth as much as we study about above the surface of the earth”. Besides the issue of hydrocarbon resources, recovery of geothermal resources, subsurface storage of nuclear waste, subsurface storage of CO_2 from the biosphere, recovery of rare earth minerals, and compressed air storage are all areas that strongly need subsurface characterization and studies. As such, the educational needs of professionals working in these fields requires inclusion of many subsurface technologies.

Fairhurst (2017) also indicates that most of the research and development investments for subsurface studies in the past have been financed by companies engaged in petroleum exploration and research support in other areas has been minimal. For example, geological engineering has focused on groundwater issues and the mining industry has focused mostly on equipment and rock mechanics. As such, studies of subsurface aquifers and subsurface reservoirs, as storage space, can also greatly benefit from some of the technologies developed for hydrocarbon reservoirs.

Pressure transient testing methods help in obtaining information on the integrity of reservoir cap rock. Weaknesses in caprocks and wellbore integrity can negatively affect the safety of fluid injection approaches such as gas storage, wastewater disposal, and subsurface CO_2 storage. But when there is a periodic flow rate pulse in the injection well and the pressure pulses travelling through the caprock and are observed in the zone above, the hydraulic characteristics of the low permeability caprock can be estimated using pressure amplitude.

It takes contributions from geophysical mapping, outcrop studies, well sample or mud log studies, wireline logs or LWD, core analysis and production logging to examine a potential subsurface reservoir as a target for production or injection purposes. What can enhance the evaluation is the information from pressure transient test.

Muskat, in 1937, examined the potential relationship that may exist between the reservoir permeability and the build rate of well pressure. Horner, in 1950 showed that

I. Ershaghi, *Solved Problems in Well Testing*,
https://doi.org/10.1007/978-3-031-47299-2_1

a plot of Pressure versus Horner time defined as $\frac{(tp+dt)}{dt}$ on a semi log paper leads to a straight-line relationship. Here, (tp) is defined as the production time and (dt) is the shut-in time. Miller et al showed that if producing time is long, in fact P versus dt can be a straight line. Transient pressure analysis techniques are recognized as excellent tools for reservoir property analysis. One can obtain formation characteristics and reservoir heterogeneities. These techniques were initially developed for radial flow conditions. For wells intercepted by hydraulic fracture the flow behavior is different than unfractured wells.

The major strengths and weaknesses of various interpretation methods have been extensively discussed in the literature. The expertise in the analysis of pressure transient tests often seems complicated with too many equations and too much mathematics. The reality is that we can breakdown the technology into qualitative and quantitative categories. The qualitative expertise that consists of specialized plots to examine patterns can be learned with minimal mathematics expertise. The patterns obtained by well tests themselves can reveal much about the condition of a subsurface reservoir.

That is the emphasis of this book. For each type of subsurface reservoir study and wellbore configurations, three sets of guidelines are included. First the focus is what test can provide in terms of beneficial information. Second, the patterns of well pressure transients versus time for carefully designed test wells is illustrated. Finally, for those interested in obtaining quantitative measures from the transient test data, some basic equations are also shown with examples and applications.

1.1 Objectives

The objectives of testing exploration or appraisal wells can be summarized as follows:

- Determining the nature of the formation fluids,
- Measuring well productivity,
- Measuring reservoir temperature and pressure,
- Obtaining samples for laboratory PVT analysis,
- Obtaining information on reservoir description (permeability, heterogeneity),
- and Estimating completion efficiency, i.e., the skin factor.

The technology of interpreting transient data well is both qualitative and quantitative. As such, we can consider this as a tool in quantitative geology. The technology of well testing includes recording fluid production or injection rates and monitoring pressure measurement signals from wellbores and relating those signals to dynamic recognition of subsurface geology and reservoir properties. It is aimed at helping geoscience professionals, students of well testing as well as practicing petroleum engineers in subsurface reservoir characterization and modeling. It is also aimed at helping engineers of various backgrounds who get involved in both the qualitative and quantitative analysis of pressure transient test data on production or injection wells. This group can also become fluent in the solution methodologies or rely on computer aided methods for parameter estimation. It offers techniques for improving skills in evaluating subsurface saline aquifers, geothermal reservoirs and gas and other liquid bearing formations. Finally, there are those who by virtue of their technical expertise can derive and develop new solutions and handle complex cases.

While still new analytical solutions are being developed and published by well test professionals, the advent of numerical well testing has opened new frontiers for modeling and handling complex cases that otherwise cannot be solved using single well analytical solutions.

The book includes 17 chapters. After Chap. 1, the Introduction, the focus in Chap. 2 is the premise of well testing including the opportunities and the type of information that can be generated. This Chapter offers a summary of general interpretation methods that can integrate and include the training needs of various groups of professionals. In Chap. 3, the qualitative aspect of well testing with a focus on pattern recognition of well test data is discussed. In Chap. 4, the focus is parameter estimation in which issues of estimation of various reservoir parameters are reviewed. In Chap. 5, the issues of handling gas and gas condensate reservoirs will be dicsussed. There are also specific methods related to naturally fractured reservoirs that will be discussed in Chap. 6. Many wells are drilled and completed with hydraulic fractures. This subject will be the focus in Chap. 7. Also, testing of horizontal wells that increase contact area, especially in tight rocks, will be discussed in Chap. 8. There is interwell information that can be obtained from pulse and interference tests. That will be the subject of discussion in Chap. 9. In dealing with unconventional resources, where rate measurements are primarily the basis of evaluation, characterization of the fractures and pressure interference is an important topic that will be discussed in Chap. 10. When it gets to evaluating the performance of injection wells, much can be inferred from well testing. This is a topic that is the focus of Chap. 11. There are specific signals that can be obtained when testing subsurface formation with the application of drill stem tests. This topic will be discussed in Chap. 12. Over the years, significant progress has been made in the use of computers in well testing. Among these data de-noising, application of artificial intelligence (AI) and numerical well testing are significant. These topics will be discussed in Chap. 13. We introduce matters related to pumping well issues in Chap. 14. Also, well tests to estimate various parameters require proper design. Topics related to test duration and design will be reviewed in Chap. 15. In Chap. 16, the application of well testing to groundwater hydrology is briefly discussed. Finally in Chap. 17, a set of problems and their solutions are included that cover a summary of patterns related to well-test data.

References

Fairhurst, C.: Earth resources engineering. Hydraul. Fract. J. **4**(1), 9–16 (2017)

Horner, D.R.: Pressure buildup in wells. In: Third World Petroleum Congress. Leiden, The Netherland, Section II, Preprint 7, pp. 25–43. (1951)

Miller, C.C., Dyes, A.B., Hutchinson: Estimation of permeability and bottomhole pressure from bottomhole pressure buildup characteristics. JPT 91–104 (1950)

Muskat, M.: Use of data oil the build-up of bottom-hole pressures. Trans. AIME **123**(01), 44–48 (1937). https://doi.org/10.2118/937044-G

Chapter 2
An Overview of Well Testing

Pressure transient analysis is a powerful tool that can assist in characterizing subsurface oil and gas reservoirs. The analysis can provide important insights into the nature of reservoir heterogeneity. Understanding heterogeneity can facilitate the understanding of the reservoir dynamics and can help in the efficient development of complex subsurface reservoirs. Knowledge of heterogeneity plays a significant role in any reservoir mapping and development of such reservoirs. Well tests not only can characterize the nature of reservoir boundaries, but they can also help to identify fault zones and estimate fault permeabilities. These measurements can also help in mapping reservoir properties with respect to the distance from a well and can identify fluid–fluid boundaries, vertical communication among layers, and identify drive mechanisms such as gas cap drive, water drive, solution gas drive and gravity drainage.

The use of well tests requires information on professionally designed tests, adequate testing time, accurate measurements and understanding the importance of consulting other sources of information. It also includes justifications such as the cost-effectiveness of information value. From the standpoint of assessing a subsurface reservoir as a source of production or for a place of injection and for using it as a storage site, there is a need for the well paths to reach those structures. We assess these wells, and we can equip our analysis based on parameters that are under our control. These include the duration of the tests, measurement parameters, and understanding of the patterns observed in the data.

The primary measurements are first pressure changes with time and the production and injection rate changes with time. How we represent these trends is all based on solid formulations that have been developed and presented in the pressure transient testing literature. The purpose of this book is not to repeat such formulations that can be found in all classic books and publications on pressure transient tests, but the goal is to introduce the transient test as a tool and offer possibilities in the form of patterns and expectations to help a subsurface geologist and engineer with new reservoir properties and confirm information from other sources of data.

I. Ershaghi, *Solved Problems in Well Testing*,
https://doi.org/10.1007/978-3-031-47299-2_2

Well testing is a process that by creating an input signal, that usually includes a sudden change in flow rate, we measure the pressure change with time. It is one of the essential sources of dynamic data that can help to characterize various reservoir properties, heterogeneities, reservoir boundaries, and flow connectivity. Using pressure transient combined with borehole imagery and seismic data can all help to better characterize subsurface reservoirs.

One advantage of running well tests is to recognize the existence of heterogeneity. These could relate to the presence of channels, hard streaks, and fractures that can be recognized by well tests. Such heterogeneities may lead to the estimation of higher or lower permeabilities compared with core-log permeability as core analysis tends to overlook high permeability areas.

Permanent downhole gauges can help to measure real-time bottom-hole pressures and temperatures. A cautionary note is that these gauges may fail in high-pressure/elevated temperature conditions unless the measurement tools can manage extreme temperature and pressure conditions.

Fractures may include diffuse fracture networks and fracture corridors. Diffuse fracture networks improve average permeability, and fracture corridors act like high conductivity flow channels deep into the reservoir. Recognition of the presence of fractures is important as fractures can have first-order controls on fluid flow in subsurface reservoirs. From well test data, fractures can be characterized in terms of aperture, density, distribution, conductivity, and connectivity. In general, well-testing can significantly help in fracture characterization for comparison with outcrop analogs.

Pressure transient tests are also useful in characterization of coal seam reservoirs. Changes in coal samples, from their native subsurface state can be significant. Variations such as gas desorption, and deformations can change the character of the core samples.

2.1 Characteristics of Subsurface Reservoirs

Starting with the oil and gas deposits, there are some geologic structures that are called classic petroleum traps. Examples include structural traps such as anticlines, fold traps, fault traps, and salt piercements. Other types include permeability barriers such as pinch-outs and channel deposits.

Subsurface reservoirs can be classified from the geological point of view or the fluids they contain. Geologically, subsurface hydrocarbon reservoirs may consist of tight or high permeability sandstones and or carbonate rocks. Typically, the fluids in subsurface reservoirs may be water or water plus hydrocarbon liquids and hydrocarbon gases. Hydrocarbons can include oil, gas-condensate, and dry gas, including heavy oil and tar sands. These reservoirs may also include coalbed methane (CBM) and gas hydrates. With the advances in completion technology, especially fracturing, even source rocks are also now included in the category of tight petroleum reservoirs or unconventional reservoirs.

Subsurface reservoirs may sometimes be formed because of the folding of the rock layers, often resulting in dome-shaped structures and anticlines, Fig. 2.1. A subsurface reservoir develops if there is a sealing cap rock. The reservoir may also be formed in fault blocks caused by shearing and offsetting of the layers resulting in the geological placement of nonporous and non-permeable rocks opposite porous formations. There are also cases where geologically, the upward movement of an impermeable salt dome deforms the geologic structure and may take the shape of a dome, Fig. 2.2.

We need to review the genesis of subsurface reservoirs. Geologic structures comprising subsurface reservoirs are ancient sedimentary systems that were deposited under various geologic sets of conditions. These reservoirs may contain salt water, fresh water, natural gas, oil, and a combination of these fluids. Based on a long history of exploration, the ones containing hydrocarbons may be source or reservoir rocks. Recovering hydrocarbons from source rocks is the basis for developing unconventional resources.

Reservoir rocks are the targets for the exploration of conventional hydrocarbon resources. Conventional reservoirs can turn into storage sites for natural gas or carbon dioxide. Many communities and utilities depend on gas storage operations. But there is now awareness of using the massive pore spaces of saline aquifers for subsurface storage of carbon dioxide or compressed air (Krishna et al. 2022).

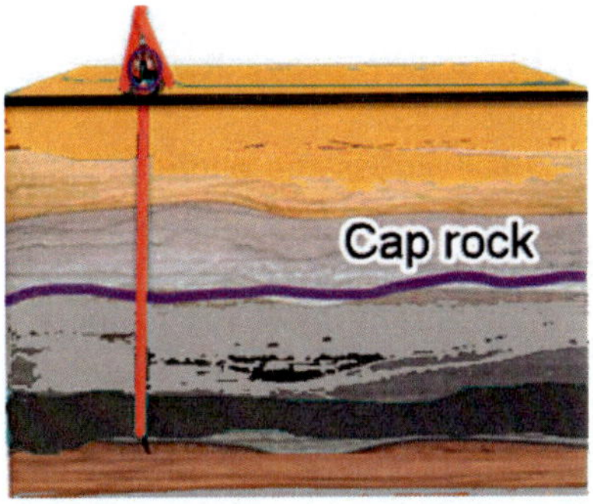

Fig. 2.1 Typical layering in subsurface reservoirs

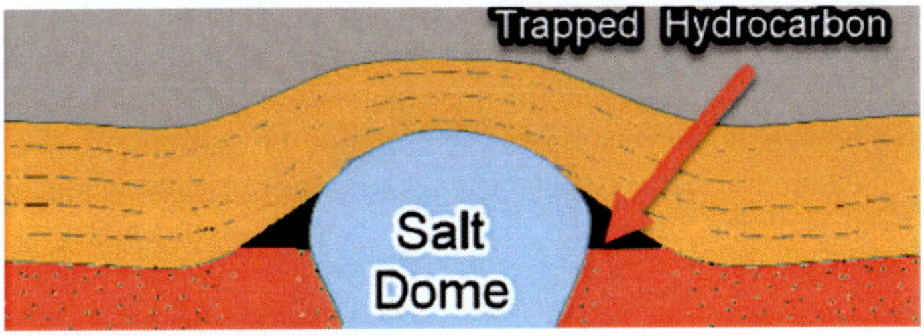

Fig. 2.2 Tarps formed by a salt dome

From the standpoint of characterization and the screening process, several qualifications must be there to designate a geologic structure as a reservoir or a subsurface storage site. This includes the reservoir extent, pore space capacity, competency of caprock, proximity to the geologic faults, permeability of faults, presence of fractures, zonation, lateral heterogeneities, near-wellbore formation damage, and permeability directions.

Permeability is an important petrophysical property of subsurface reservoirs. When measured during core analysis using air or nitrogen, the measurement corresponds to only a limited sampling size of the reservoir. However, the permeability from a well test represents the average effective permeability of the reservoir and its fluid type within the radius investigated for that test.

As indicated before, Muskat, in 1937, indicated that there should be a relationship between pressure recovery after production from a well and the reservoir permeability. He noticed that if the rise in the bottom-hole pressure or fluid level is rapid, the permeability of the sand about the wellbore is large, and conversely, a conclusion may be drawn if the rise is slow. However, Muskat realized that a more quantitative estimate could give additional value information. Kolb in 1960 described a project at Shell Development Co.'s Exploration and Production Research on the improvement of the accuracy and the ease of BHP (bottomhole pressure) measurements. Millikan and Sidwell (1931) described various simple methods for the estimation of bottom-hole pressures. Browns and Miller (1961) discussed the use of a graphical method that can help in the determination of the corrections which must be applied to the spot readings to estimate the static pressure. Rasa and Katz (1945) discussed pressure measurements related to gas wells.

Various subsequent contributions and publications have helped to establish the basis for understanding the pressure transient behavior of a well, the type of flow system, and the drainage area around a well.

Understanding reservoir anisotropy helps to predict production performance. Petroleum reservoirs can be very heterogenous, and their petrophysical properties can vary with distance from a well. What primarily defines the nature of rock deposition is the applicable facies model that can explain the character of a given subsurface geologic formation. When dealing with sediments that contain resources such as oil and gas, the expectation is that during the geological ages, the reservoir has become preserved. Otherwise, the resources may escape to other layers and even reach the surface. That protection is through the existence of caprocks and base rocks. While the vertical variation can be mapped by 2D-3D seismic imaging, the most effective tool to map heterogeneity is running tracer surveys or pressure transient well tests.

As this book is about running and interpreting pressure transient tests, we will examine pressure transient tests that can identify the presence and the magnitude of heterogeneities, Fig. 2.3 shows an example of lateral heterogeneity.

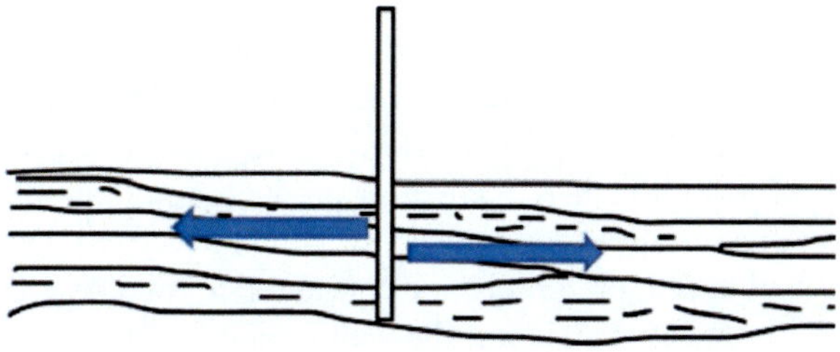

Fig. 2.3 Typical lateral heterogeneity around a well

2.2 Well Testing Technology

In the science and technology of formation evaluation, we obtain information from mud logs, wireline logs, core analysis, and seismic and depositional models governing the formation of some subterranean reservoir. But these are all static-type measurements and often offer a limited depth of investigation. The measurements provided by some formation evaluation tools represent what is around the wells. Most of these measurements do not create a flow condition leading to an estimation of formation transmissibility. Then for reservoir management studies requiring simulation modeling and for implementing operations requiring a clear understanding of reservoir permeability, we often rely on correlations such as those between porosity and permeability using core analysis data. Even in the application of technologies offering a large scan of the reservoir, such as seismic cross-sections, quantitative estimates of the permeability across faults cannot be ascertained just from analysis of seismic data or analysis of geologic models. Well- tests are needed to make such independent estimations and determinations.

Rock permeability and its changes during a depletion process is a key component in understanding a reservoir. If we manage to obtain core samples, measurement of permeability made on core samples using laboratory testing has limitations. The main shortcoming relates to the changes in the rock properties when it is subjected to conditions when core samples leave the reservoir and or if there are geomechanical changes that take place during coring or core preparation.

Well-tests, on the other hand, can allow the estimation of in-situ reservoir permeability through carefully designed flow tests. Not only do these measurements provide a large depth of investigation, but lateral changes around a well can also be detected, which is an effective way to analyze wellbore conditions and map the dynamic permeability of a reservoir. Besides the estimation of formation permeability, these pressure tests can also identify large-scale heterogeneities caused by faulting, pinch-outs, layering, and in-situ fractures.

In summary, as indicated before, well-test interpretation is the process of obtaining information about a reservoir by analyzing the pressure transient response caused by a change we make in production or injection rate. Wells drilled to a subsurface reservoir can be different, and there are also several diverse ways of categorizing tests run on wells depending on what technically and economically can be implemented.

Tests may also be focused on one well or a group of wells. In single-well tests, the near-wellbore region can be examined to quantify the damage or effectiveness of a stimulation job. In a drawdown test, the reservoir is initially at a uniform pressure. The well is then opened for production, preferably at a constant flow rate, and the pressure drawdown response is monitored. In a buildup test, after a well has been produced, preferably at a constant rate for a period, a pressure gradient is created in the reservoir. Then if the well is shut in, the resulting pressure build-up response is measured, and that can help in the estimation of reservoir conditions and properties.

Single well tests may include a production period followed by a shut-down period, or they could include changing rates and observing the pressure responses. The objectives are also different whether the test is run in the early life when the focus is an assessment of the subsurface reservoir or later during the operating life when changes in wellbore conditions need to be monitored. The goal can be testing to estimate a well's deliverability and its potential. The intention also can be estimating the inflow performance relationship (IPR) curves. But the primary objectives of transient tests are to estimate the in-situ permeability to the produced or injection fluid, to evaluate average drainage volume pressure, to examining the need for stimulation treatment, assess the effectiveness of stimulation jobs, and finally for estimating fluids in place within the drainage area of wells.

If the goals are to quantify the degree of communication among wells or estimate directional permeability, multi-well tests are run. Changing the production or injection rate in one well generates transients in the nearby wells. Much can be realized from such tests in terms of estimating reservoir heterogeneities and reservoir anisotropy.

Objectives of an exploration well test usually include fluid sampling, estimating initial reservoir pressure, evaluating well productivity, estimating distances to boundaries, and estimating fluids in place. Exploration well tests are often conducted while the rig is on location, making such long-term tests extremely expensive. The expense is often justified by the major investment decisions that will be made based on the information obtained from the test.

In contrast, in development well testing, the focus is on the individual wells and the near-wellbore area. Objectives of a well test on a development well can also include estimating average drainage area pressure, evaluating effectiveness of stimulation treatment, quantifying wellbore damage, and for estimating reservoir permeability. Compared to exploration well tests, development well tests are less expensive. The results of development well tests are used in investment decisions, such as whether to do a stimulation treatment.

In summary, well-test interpretation can play a key role in many stages of the life of a well, including exploration, reservoir development, and production engineering.

2.3 Problem Statement

In this book, we discuss well-transient testing methods and the host of information one could obtain from such tests. Some of the typical questions that can be answered from well test include the following:

- Why is a given producing well underperforming?
- Can the production or injection performance be improved?
- Are several wells communicating with each other?
- Where are the fluid flow boundaries, their characteristics, and the role of geological boundaries?
- Is there any evidence pointing to the presence of natural fractures?
- Is there any evidence that a hydraulic fracture has been formed?
- Are there reservoir heterogeneities or anisotropies around producing wells?
- What are the effective flow regimes around a given well?
- Could we obtain real-time petrophysical properties?

Permeability in underground structures can vary substantially. If permeability is decreased to zero, the consequence will be sealing off a portion of the reservoir or causing a permeability pinch-out. During a field shut-in condition, it may take an exceptionally long time to equilibrate the pressure in the reservoir rock. The structural seal in subsurface geologic characteristics can mean a constant volume. An impermeable layer may overlay the porous geologic body and is called a cap rock. Caprocks can prevent the vertical or sometimes the lateral movement of the fluid to outside the geologic reservoir. The presence of a lateral seal caused by a geologic fault may also produce a leak at one or more sides of the subsurface reservoir.

Fluid flow around faults can be complex. Once such a deformation-related effect exists, we can expect changes in directional permeabilities. The boundary at the bottom of the porous medium may also be sealed off by some impermeable rock or by constant bottom water. Faulting of subsurface formation happens in response to situations that sufficiently stress lithologies and the resistance at the time of stress termination. Estimation of fault permeability is within the capabilities of well transient tests.

Burial depth is more of a local influence with rocks and relates to the position of rocks as discovered during exploration activities. Also, rock compressibility is affected by the continuous burial of sediment and impact of load on rocks, and the impression of principal stresses.

Anisotropy of permeability exists when the measured permeability varies depending on the direction of fluid flow through the reservoir. Permeability anisotropy can also be caused by orientation during deposition and deformation or through the preferential orientation of pore connectivity or porosity from open microfractures during deformation. Reduction in permeability can occur due to grain size rearrangement, compaction, and cementation. During an injection phase, fluid is injected into the reservoir, the increased pressure causes compression of exiting in-situ fluid and can affect the permeability.

The reality is that the subsurface reservoirs are heterogeneous. That can be part of the reason that two producing wells do not behave the same way, or two injection wells do not take the same amount of injection fluid. The main question is, then, what are the causes of subsurface reservoir heterogeneities? The answer relates in part to realizing the nature of rocks comprising reservoir structure and the geological processes that resulted in the formation of a particular reservoir. In terms of the constituents, subsurface petroleum reservoirs are primarily sedimentary rocks consisting of sandstones, conglomerates, carbonates, and shales. Measurements from drilling and subsurface imaging have shown drastic variations both vertically in layering and laterally in sedimentation. Images generated from well measurements and interpolations from various techniques have shown variations in available pore volume, horizontal and directional permeabilities, and capillary properties affecting fluid distributions.

During the last several decades, much effort has been focused on building reservoir cellular models. The main question from the standpoint of a subsurface geologist can be whether these images represent the geological processes resulting in the formation of such reservoirs.

Before discussing the benefits of using information from well-tested responses, we must review important areas for understanding and monitoring subsurface reservoirs. These can relate to the original deposition, changes caused by production-related pressure depletion, and changes caused by pressurization from fluid injection.

Variation in the performance of individual wells can be explained by lateral heterogeneity. Even vertical accumulation of sediments represents different geologic times and facies models. We know this because of studies that have related outcrops to subsurface systems, Myall (2022). Relating the structure of subsurface reservoirs using facies or depositional models is a clever way to answer some of these questions. In the language of facies models, subsurface reservoirs are formed under one or a combination of geological processes. The net result is in systems such as alluvial fans, fluvial deposits, eolian deposits, deltas, near-shore deposits, lacustrine deposits, turbidite fans, and others. Some of these are considered continental deposits. Some are marine deposits, and some are considered as mixed deposits. A good example is considering braided and meandering river deposits. The nature of these deposits can explain the lateral variation in rock properties.

Our limited access to subsurface rocks is via wells drilled and subsequent sampling from such wells. When defining and characterizing a subsurface reservoir, we consider the estimation of porosity and permeability, and boundaries to be the essential measurements. Porosity defines pore space availability, and permeability is related to the ability of the rock to transmit fluids. Boundaries define the extent of a reservoir. In the description of rock for storage purposes, static measurements such as logs, and core analysis data can be informative. The greatest task, however, is estimating permeability. Measurements of permeability from sidewall or full cores have limited value as such data relates to a small portion of the rock body. Scanning larger volumes is difficult as permeability can also have variations in different directions of a subsurface reservoir.

Anisotropy of permeability exists when the permeability varies depending on the direction of fluid flow through the reservoir. Permeability anisotropy in clastic rocks can exist because of orientation during deposition and deformation or through the preferential orientation of pore connectivity or porosity from open microfractures during deformation. In carbonate rocks, there are heterogeneity effects that control permeability trends. Besides the original sedimentation-related causes, reductions in permeability can also occur due to grain size rearrangements, pore size reductions and fluid-related precipitation and cementations. Fluid flow around faults can be complex. Once such deformation-related effects exist, we can expect changes in directional permeabilities that can affect resource recovery, injection efficiency, and deliverability.

Dynamic processes associated with depletion and movements in the crust, including the impact of tectonics and compaction, tend to change the effective permeability of rock bodies. There is a need to look at ways to get a realistic and dynamic view of existing and developed lateral permeability changes. This is important in terms of resource recovery of hydrocarbons, geothermal fluids, and subsurface water resources. It is also essential for evaluating suitable underground carbon capture and underground storage (CCUS) sites and for storage and deliverability aspects of compressed gases for gas storage and compressed air for energy storage.

There are several reasons that such variations exist in terms of the ranges of permeability values and lateral changes in permeabilities. Some are related to the depositional model that can describe sedimentation. For example, for fluvial deposits, as shown in Fig. 2.4, there can be several orders of magnitude changes in permeabilities with directions. The direction of earth stresses can also impact the permeability trends.

Additionally, as shown in Fig. 2.5, the presence of reservoir fractures can result in directional changes in the rock body permeability.

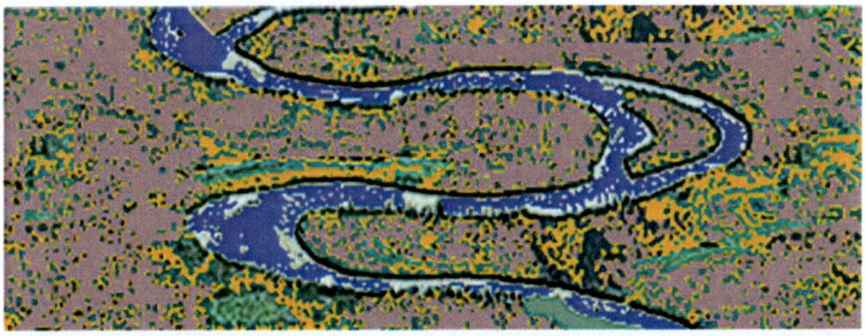

Fig. 2.4 Heterogeneities in a meandering river depositional environment

Fig. 2.5 Heterogeneities caused by the presence of fractures

Many localities are also interested in developing gas storage fields. Conversion of existing semi-depleted oil and gas fields to gas storage requires studies on caprock integrity and the geologic formation storage capacity and deliverability. There is also a growing need for sustainable deep underground geologic formations for the storage of carbon dioxide. Both onshore and offshore subsurface reservoirs have been targets of interest.

At one time, the focus of subsurface reservoir management study was to look for hydrocarbon or geothermal resources or to use semi-depleted reservoirs for gas storage operations. This goal is there as we still need hydrocarbon-based fuels or geothermal fluids to run economies. There are, however, important developments in the use of subsurface pore spaces in the form of semi-depleted hydrocarbon reservoirs or saline aquifers for storage that will require constant and reliable pressure monitoring.

With the emphasis on low-carbon engineering, CO_2 storage in aquifers is an attractive option for reducing CO_2 emissions into the atmosphere. But subsurface storage of carbon dioxide (CO_2) introduces geologic and economic uncertainties. The main issue of CO2-subsurface-storage is a clear understanding of the subsurface geologic heterogeneity. As mentioned before, the subsurface reservoirs for CCUS include saline aquifers, semi-depleted hydrocarbon reservoirs, deep coal seams, and organic-rich source rocks. Carbon dioxide could also potentially be stored in volcanic rocks such as basalts because of the mineral trapping capabilities. In deep geologic structures, if the temperature exceeds 88 F and pressure exceeds about 1057 PSI, CO_2 stores as a supercritical fluid. This means substantial volumes of CO_2 can be compressed into a smaller volume.

With the great storage spaces that subsurface reservoirs and aquifers offer for energy storage and GHG sequestration operations, what is important to establish is the volumetric capacity, seal competency, and deliverability characteristics of such subsurface reservoirs. Volumetric capacity is represented by rock porosity and the rock bulk volume. Deliverability estimation is a function of rock permeability and formation thickness. Integrity depends on the caprock and geologic faulting systems.

Toward a carbon–neutral economy, well-testing concepts are needed and quite relevant. For example, in subsurface storage of carbon dioxide, well-testing methods can play a key role in evaluating the competency characterization of caprock and in permanent leakage monitoring of subsurface storage sites. Furthermore, the evolving need to use subterranean reservoirs as gas storage or saline aquifers for compressed air storage for excess energy generated from renewable sources such as solar and wind requires continuous reservoir pressure monitoring through well testing, and interpretation. As such, in managing conventional and unconventional reservoirs, we see an increase in the application of well-testing concepts. As indicated, while there is interest in locating super-sized subsurface geologic storage sites to dispose of greenhouse gases, the interest is also growing in building subsurface storage for compressed air to store energy from renewables or to manage excess energy from gridlines.

Among the factors to consider, subsurface geological mapping is the first step in characterization effort. But as geo-stresses change and major alterations develop when subsurface rocks are exposed to injected fluids, there is a definite need to include in the information that can be generated from dynamic tests and measurements, well transient tests.

Rock anisotropy can manifest itself with lateral changes during sedimentation and various levels of sediment lithification. In contrast with brittle rocks, shallow and mostly unconsolidated sedimentary formations usually can experience substantial deformations. Some of the injection operations, such as CO_2 disposal, can also cause changes within the reservoir and caprock.

Storage capacity is the space within a certain porous geological structure or reservoir for storing fluid. With injection, there will be a rise in the reservoir pressure. The maximum amount of fluid stored in a geological system depends on the maximum acceptable pressure increase, which will not activate existing faults and/or form new cracks.

There are many ways that storage of waste fluid can take place, including structural trapping, residual trapping, solubility trapping, and mineral trapping. The issue of concern for the purpose of this discussion is structural and residual trapping. In structural-type trapping, the sealing capability of caprocks helps to prevent leaks to the surface. Also, if there are natural fractures or high permeability channels, communication with faulting in the area can raise concerns. One way to monitor the real-time changes in fluid storage is continuously monitoring reservoir pressure.

Waste fluids or gases may be stored in several diverse ways. The storage is primarily held in underground under pressure in three types of storage sites. This includes depleted reservoirs in oil and/or natural gas fields, aquifers, and salt cavern formations. Each storage type has its own physical characteristics (porosity, permeability, retention capability) and economic factors (site preparation and maintenance costs, deliverability rates, and cycling capability), which govern the suitability for applications. Depleted oil and natural gas reservoirs are the most used underground storage sites because of their wide availability. Most existing natural gas storage is formerly depleted natural gas or oil fields close to consumption centers. A field conversion from production to storage duty takes advantage of existing wells, gathering systems, and pipeline connections.

Two important characteristics of an underground gas storage reservoir are the capacity to hold natural gas for future use and the rate at which gas inventory can be withdrawn for managing the deliverability rate.

In some areas, saline aquifers have been converted to natural gas storage reservoirs. An aquifer is suitable for gas storage if the water-bearing sedimentary rock formation is overlaid with an impermeable cap rock. Although the geology of aquifers is like depleted hydrocarbon producing fields, their use for natural gas storage usually requires more cushion gas, which means less flexibility in injecting and withdrawing. Deliverability rates may be enhanced by an active water drive, which can support the reservoir pressure through the production cycles. Worldwide, saline formations have the largest potential volume for storing fluid or energy using compressed air. Extensive saline formations exist in the large sedimentary basins located across the world. For fluid storage operations in subsurface structures, storage space, injectivity, close boundaries, and depth are all essential parameters. Injectivity refers to the rate at which any fluid can be injected into the subsurface and is essential as it relates to formation permeability. A storage complex must have one or more confining zones that seal above the injected formation that are intact and do not have leakage pathways.

Salt caverns provide extremely high withdrawal and injection rates relative to their working gas capacity. Base gas requirements are low. In the US, most salt cavern storage facilities have been developed in salt dome formations in the Gulf Coast states. Cavern construction can be more costly than depleted field conversions. Still, the ability to perform several withdrawal and injection cycles each year reduces the per-unit cost of each thousand cubic feet of gas injected and withdrawn.

A major cause of lateral heterogeneity is compartmentalization. Permeability barriers within a field cause compartmentalization which is the geological segmentation of reservoirs into isolated compartments. The combination of stratigraphic architecture, structural architecture, fault permeability, and diagenesis affects the formation of compartments.

Understanding the stratigraphic and structural variations in permeability is essential to map compartments. For example, fluvial and fluvial-deltaic sediments can be complex to characterize. In massive sands, heterogeneities are caused by lithic fragments and differential bioturbation; in bedded sands and the presence of dispersed clays may be a cause for heterogeneity. Fault properties can constrain structural compartmentalization. Boundary seals constitute the barrier against the migration of fluids in geological storage projects. Bounding seals are composed of an overburden immediately above the injection horizon and under the burden and the wellbore systems used to access the injection horizon. Changes in the integrity of the subsurface can have significant economic and environmental consequences. Understanding these changes helps to mitigate the development risks of subsurface reservoirs and optimal development of the field.

In actual subsurface geologic reservoirs, the position of geological and fluid boundaries can vary in all directions. There is a need to map these boundaries. The dip calculations from density and gamma-ray images represent areas close to the wellbore and only represent a small reservoir volume. We could extend the imaging with ultradeep electromagnetic (EM) technology, where changes can be identified more than 100 ft. from a borehole. This still only provides a limited image of the reservoir.

As such, a deep representation of lithology and fluid boundaries using various tools is difficult because the images do not directly indicate a distance to the resistivity boundaries identified. The field development plan needs integration of geophysical, geological, petrophysical, and reservoir engineering, including well-test data. While among all the long investigating tools, high-quality seismic data can help to unravel the complexities of subsurface geology, well tests data can help significantly in calibrating the seismic signals related to lateral changes, compartmentalization, and dynamic properties. Advances in using permanent gauges also allow monitoring of wells whether producing, idle, or abandoned. Again, well-testing methods play a key role in continuously interpreting real-time sensor data and monitoring changes in subsurface storage systems.

In summary, well testing techniques can help the non-experts to improve skills in the interpretation of well tests, and it can offer capabilities for understanding and forecasting of production from aquifers, gas fields, oil fields, geothermal fields, storage fields and wells-producing from unconventional resources.

2.4 Questions and Answers

Q2-1: The rate in Well A producing with a rod pump has declined to values below the economic limit. Other wells in the reservoir are doing well. How can information from well-testing help to ascertain the cause of low productivity, Fig. 2.6?

A2-1: One main cause for the loss of productivity is perforation plugging or formation damage caused by the buildup of damage in part related to asphaltene buildup or clay swelling. A pressure buildup test measuring the pressure changes with acoustic signals can identify this cause, and then remedial actions may be taken to address the wellbore productivity issue.

Q2-2: A well produces from a gas condensate reservoir. Gas productivity has declined rapidly, Fig. 2.7, while indications are that the reservoir pressure has stayed high. What could cause this condition?

A2-2: A well test can detect the formation of a condensate zone around the well. This can be why starting a cycling operation requires injecting dry gas and keeping the pressure around the well above the dew point pressure until most or all condensates are recovered.

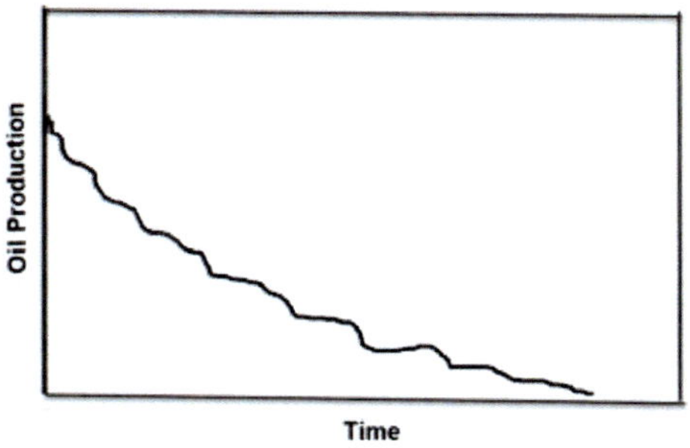

Fig. 2.6 Example well productivity decline

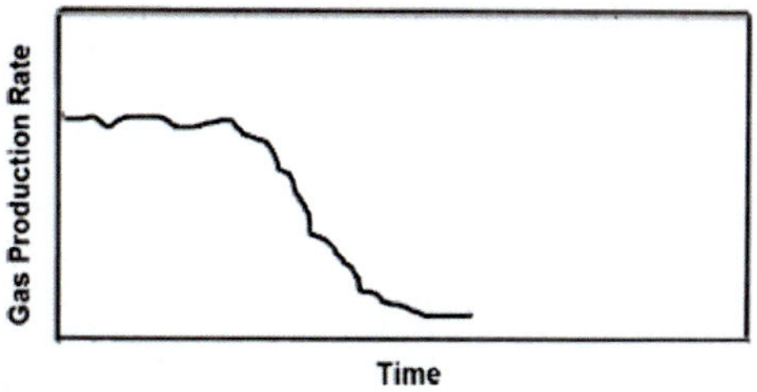

Fig. 2.7 Production decline caused by condensate banking

Q2-3: A seismic cross-section identifies the presence of a fault, Fig. 2.8. How can we establish transmissibility across the fault?

A2-3: A professionally designed pulse or interference pressure test between two wells across the fault or just a pressure buildup at one of the wells can help to establish independently the nature of fault transmissibility.

Q2-4: In a water injection well, Fig. 2.9. The concern is whether the injection gradient has fractured the intervals.

A2-4: A pressure fall-off test can establish the flow regime and independently evaluate the formation condition. If the radial flow is persistent from the beginning, the well has not been fractured.

Q2-5: A new gas field has been discovered by drilling a few wells. How can maximum deliverability be estimated for the field?

A2-5: A series of back pressure tests can help to establish the absolute open flow potential of the wells by plotting (Pr^2–Pwf^2) vs. q, the gas rate, on a log–Log paper, and the maximum deliverability, which can be estimated by using atmospheric pressure as the wellbore flowing pressure, Fig. 2.10.

Q2-6: How can the formation of a gas cap be established from well test?

A2-6: A plot of Pressure versus log(t) may identify dampening changes caused by the presence of a gas cap manifested by a drop in the slope, Fig. 2.11.

Q2-7: How can a pressure buildup test point out the presence of a naturally fractured reservoir?

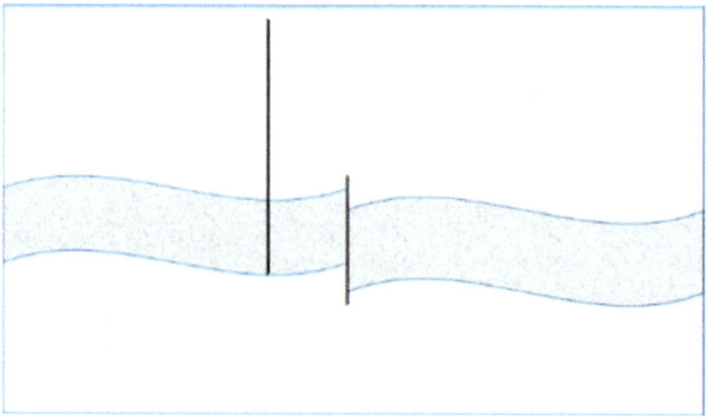

Fig. 2.8 Fault planes seen on a seismic cross section

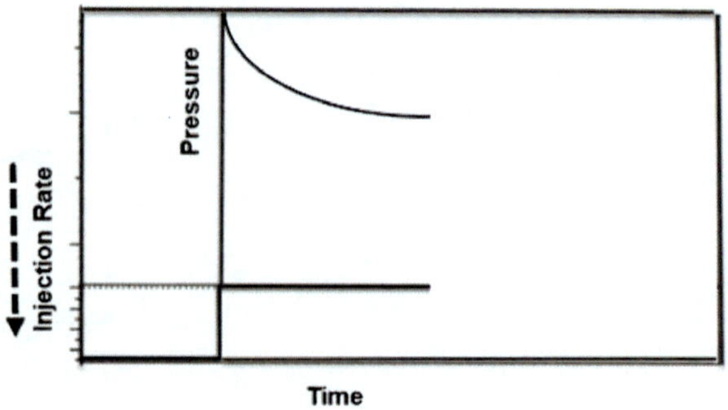

Fig. 2.9 Typical response in a pressure fall-off test

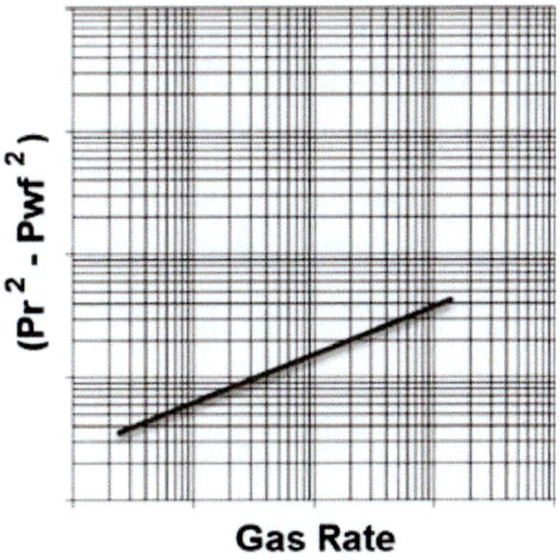

Fig. 2.10 Gas deliverability testing

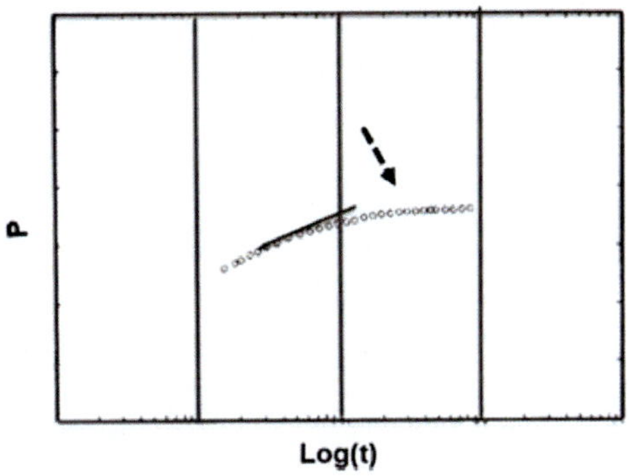

Fig. 2.11 Gas cap effect on pressure build up test

A2-7: Naturally, fractured reservoirs show distinctive patterns on the graphical representation of well test data on a semi-log plot of Pressure vs. the Horner time. This is only possible if the primary porosity is enhanced by microfractures. Otherwise, the typical short duration of the well test may be insufficient to allow observation of a dual porosity system, Fig. 2.12.

Q2-8: A horizontal well has been drilled. But it needs to be clarified if the effective length of formation contact is what is indicated in the drilling record. Could a well test resolve the issue?

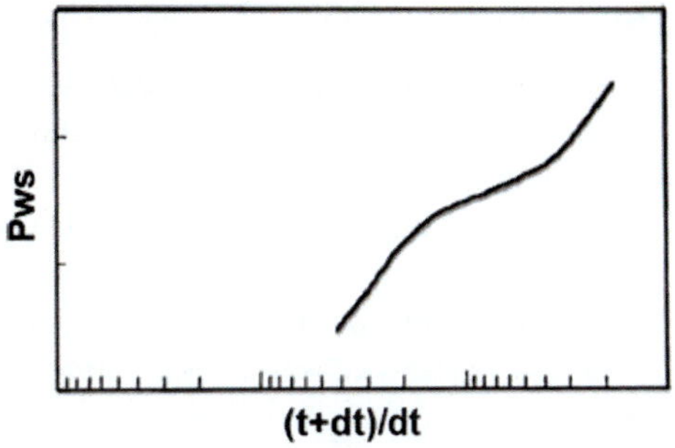

Fig. 2.12 Typical semi-log horner plot of pressure transient response in a naturally fractures reservoir

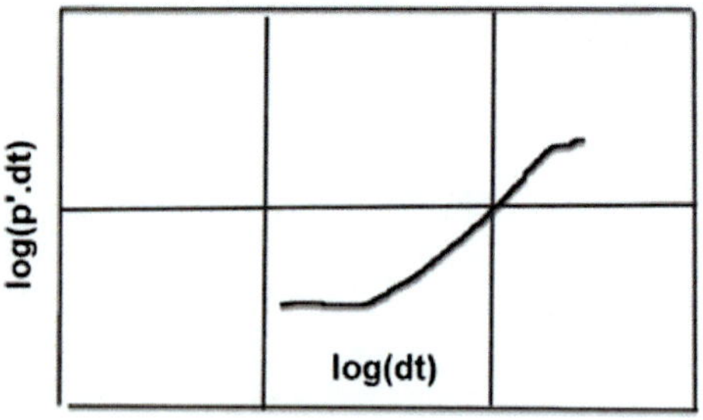

Fig. 2.13 Pressure transient analysis of a horizontal well

A2-8: Potentially well test data can help in estimating the effective lateral length of a horizontal well using an iterative process, Fig. 2.13.

Q2-9: There is concern about whether a water well is excessively draining the aquifer shared by an adjacent landowner. Could this be verified?

A2-9: Well transient tests can identify the radius of influence for water wells.

Q2-10: How extensive is a newly developed hydraulic fracture?

A2-10: Well tests can help to establish the extent of a developed fracture.

The above examples are just a few opportunities that well test solutions provide to help in facilitating asset management. But there is more. In the sections below, we discuss more additional applications of well tests.

To start with, there are a few principles we need to review. These principles are based on testing the behavior of wells with respect to idealized conditions. Reservoirs and wellbore conditions are often complicated. Yet, when we compare their responses to idealized models, we learn whether a lumped parameter approach can resemble the actual well behavior to idealized conditions. For example, in an idealized condition, where a well penetrates a single layer, it is easy to look at the expected relationship that describes how pressure transients change with time. The theory shows that in such an idealized condition, assuming the fluid flow is governed by a radial flow system, graphically, a plot of pressure vs. time on a semi-log paper will result in a straight line. Then for an actual well test, if we see the same plot and, by coincidence, a straight line is formed, we may be tempted to label that test as a radial flow condition. We then use formulations developed for idealized systems and estimate formation properties.

When we work with actual data, we may observe deviations from the expected trends and relationships, and those deviations each may have an explanation. Some relate to wellbore-dominated effects. The skin around the wellbore may also contribute to this effect. Some relate to lateral changes in fluid transmissibility, some indicate the effect of layering, and some relate to boundary effects.

Running a well-test is costly, but of course, the information obtained is unbelievably valuable. The challenge is identifying fit-for-purpose tests. What makes test interpretations easier are using sensitive gauges, assuring sufficient duration of the tests, competencies in selecting models, and priorities in the use of the tests. For example, if the goal is to identify whether a pumping well is suffering from wellbore skin, we do not have to conduct a long-duration test or need overly sensitive gauges.

In summary, to make sense of actual data plots, one needs to recognize the idealized patterns associated with any of such effects. This is an important component of well-test interpretation competency that requires knowing the expected patterns and recognizing pattern changes seen on actual tests. This capability is important because it simplifies the task of reservoir characterization. When well interference or pulse tests are conducted between two or more wells, many more capabilities are available to examine the deep characteristics of internal reservoir structure and fluid boundaries.

Mastering the technology of well test design and interpretation requires an understanding of idealized responses, recognition of well test patterns, and relationships that can allow estimation of wellbore and reservoir properties.

2.5 Expertise Needed

Subsurface reservoir engineers are solution providers. In every profession, there are tools and technologies that assist the professional in doing their work. What affects the results is the nature of the tools and techniques of subsurface reservoir characterization and engineering. We also need to incorporate information from various sources of measurement to study a particular subsurface reservoir. Other sources of information may include:

- wireline or LWD Logs,
- core analysis,
- geophysical Information,
- performance data,
- tracer surveys,
- geomechanical data,
- facies modeling,
- geochemical data,
- inverse modeling, and
- well test data.

We can group information obtained from these sources into two categories. One is static, and the other is dynamic. As indicated before, well-log data, core analysis, 3-D geophysical data, and facies modeling all represent measurements at a point in time. For example, one may estimate hydrocarbon saturation using well logs. But that type of measurement, when calibrated properly, shows the measurement at that given point in time. We can call these static measurements. These are considered static because, in most cases, there are usually no permanently installed tools to obtain the information that change with time. Unless there is the intention to run a 4-D seismic, the conventional seismic measurements also represent just one image of the subsurface. If there are dynamic changes that affect the reservoir, a fixed point in time measurement will not represent the changes. Besides well tests, tracer and performance data also provide dynamic type information. Measurements of the reservoir and near wellbore conditions with permanently installed gauges or techniques such as tracer detection can give a dynamic view of changing conditions. Similarly, the measurement of performance data is also a representation of the changing nature of rock and fluid saturation properties. Using pressure transient combined with borehole imagery, seismic profiling, and sonic logs can help to characterize heterogeneity. Permanent downhole gauges can help with the measurement of real-time bottomhole pressures and temperatures. A cautionary note is realizing the opportunity and using gauges that will not fail in high pressure/elevated temperature conditions.

Professionals with expertise in well-testing fall into four categories. First, there is that group who are aware of the potential application of these tests and actively use the derived information. Subsurface geoscientists have a better view of the complexities of subterranean reservoirs. But they may need to improve their expertise in using the technology of well testing. An independent source of information from well testing for recognizing reservoir configuration can be refreshing to their analysis. As such, the first group that can benefit from the results of analysis and calculations is the subsurface geoscientist. For this group, the important expertise relates to an understanding the facies models, causes of rock heterogeneity, and factors affecting the formation of boundaries.

The second group are those who appreciate the well-testing solutions and are professionals who can get inspiration from the graphical patterns of the test data and recognize various clues to the nature of the wellbore or the reservoir. This group must have received some education and training on the expected responses of various simple subsurface reservoirs systems. They can also qualitatively look at test data and recognize patterns and clues about the actual reservoir conditions.

Then comes the third group of professionals, and those are the ones who understand the patterns and are proficient is making quantitative estimations using the appropriate formulas and equations. The final group consists of research professionals competent in mathematical skills and interested in developing innovative solutions and describing situations not addressed in the literature. This book is intended for the first three groups of professionals.

Let us review what expertise is needed to assist the first three groups. The main common denominator is a clear understanding of subsurface geology. The engineering experts who get involved in subsurface characterization may need to have the proper visions of subsurface geology. A comprehensive planning in reservoir characterization is needed before fully appreciating well testing as a valuable source of dynamic well and reservoir information.

To build a robust geological model of subsurface reservoirs, it is best if we integrate the geological as well as pressure and production behavior of any well that has been drilled into mapping the structure. In brief, the dynamics of a subsurface reservoir may be realized with tests that represent repeated measurements. Well performance changes with time, and continuous recording can produce added information from well-tested data, tracer surveys, inverse modeling, and 4D seismic data.

Permeability is an important petrophysical property of subsurface reservoirs. As indicated before, when measured during core analysis using air or nitrogen, the measurement corresponds to only a limited sample size of the reservoir. However, the permeability from a well test represents the average effective permeability of the reservoir and its fluid type within the radius investigated for that test. What is measured is the transmissibility that is defined as $T = \frac{K.h}{\mu}$. . That is, to estimate permeability from well testing we also need an estimate of formation thickness (h) and fluid viscosity (μ). Incorrect estimation of thickness will affect the estimated effective permeability. Besides the well logs, PLT can help in the estimation of formation thickness. Also. proper reservoir PVT data, including fluid viscosity and compressibility, are needed to estimate the permeability.

We need tools that can recognize the existence of heterogeneity. These could relate to the presence of channels, hard streaks, and fractures that can affect the fluid flow. Core analysis with limited investigation volume may overlook high permeability areas. Recognition of the presence of fractures is important as fractures can have first-order controls on fluid flow in subsurface reservoirs.

There is a lot that we need to know about fractures. Fracture is characterized in terms of aperture, density, distribution, conductivity, and connectivity. Fractures can be diffuse fracture networks and fracture corridors. Diffuse fracture networks improve average permeability, and fracture corridors function as high-conductivity flow channels deep into the reservoir. Such heterogeneities may lead to the estimation of higher or lower permeabilities compared with core-log permeability correlation. Well tests can significantly help in fracture characterization for comparison with outcrop analogs.

Pressure transient responses, when examined with time, show characteristics, shapes, and patterns. Before we summarize these patterns, it is important that we discuss the graphical representation of the well-test data. The symbols used include (P) representing pressure, (q) and (t) respectively representing flow rate and time, respectively. Well-test observations include relationships that one expects to observe between pressure (P) at a constant rate vs. time or inverse flow rate (dp/q) changes with time. Sometimes we examine just these parameters and most of the time, we monitor the pressure derivative function which is defined as the product of pressure derivative, and the elapse time, (P.' Δt). In the case of monitoring the rate under constant pressure, the log–log derivative plot is $(\frac{1}{q})' * \Delta t$.

We examine the graphical relationships between these functional groups and time on cartesian, semi log, and log–log plots. The pattern recognition for the cases described below relies on examining these graphical representations. For example, if we flow a well at a constant rate, the expectation is that the wellbore flowing pressure will drop with time as shown on a cartesian scale as shown below, Fig. 2.14.

This type of plot may be observed in many different wellbore and reservoir situations. What provides clues that are significant in parameter estimation is examining this relationship against various classical models. These models include levels of different complexities in terms of wellbore and reservoir conditions. Thanks to the effort of early contributors, there have appeared in the literature exact mathematical relationships that describe the functionality of P or (dp/q) with time under those conditions. These mathematical relationships require examining the behavior observed on various graphical scales.

Fig. 2.14 Pressure drop with time

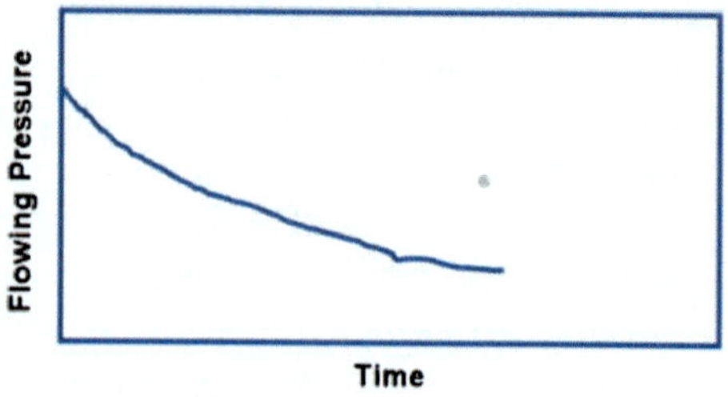

These conditions start with the simple cases of a single well in an otherwise homogeneous reservoir to cases when the formation alone cannot produce and needs stimulation treatments such as hydraulic fracturing. A general approach is to identify the type of flow regime. For example, one may start with the assumption that the flow toward a well is radial. Various diagnostic plots can help to verify the accuracy of such an assumption. Sometimes these diagnostic plots can indicate that flow regimes and reservoir conditions are different than what may have been expected by the analyst. A powerful tool in the hands of the analyst is to examine the pressure derivative function plots as the starting point. It is important to remember that many flow and reservoir conditions may produce similar diagnostic plots. Then it is important to select the most plausible interpretation by consulting other sources of information.

2.6 Pattern Recognition

The pattern recognition process for analyzing results from pressure transient tests needs special plots. These plots consist of displaying changes in the pressure with time. We use three different scales for plotting: Cartesian plots, semi log plots and log–log plots, Fig. 2.15. Cartesian plot of Pws, Pwf or ($\frac{dp}{q}$) vs. $\sqrt{t}$ can help to confirm the existence of linear flow. Cartesian plots are also helpful in displaying pressure drawdown and the manifestation of reservoir boundaries. Derivative plots are shown on a log–log scale. Semi-log scales are used to examine radial flow conditions. They can be Horner plots of pressure vs. log ($\frac{tp+dt}{dt}$) or pressure versus log(dt) if producing time is large, Miller et al. (1950).

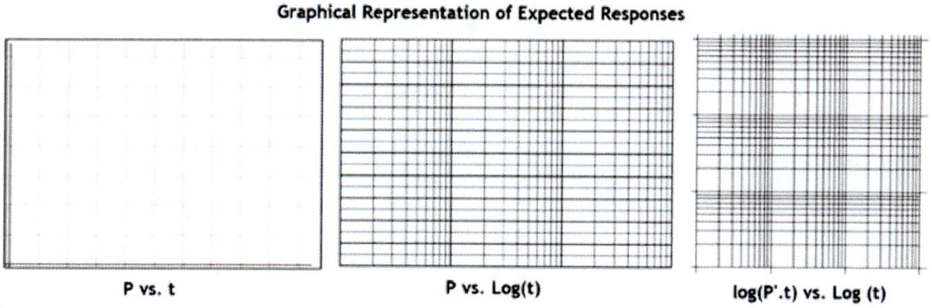

Fig. 2.15 Various graph scales

References

Horner, D.R.: Pressure buildup in wells. In: Third World Petroleum Congress. Leiden, The Netherland, Section II, Preprint 7, pp. 25–43. (1951)

Kolb, R.H.: Two-bottomhole pressure instruments providing automatic surface recording, JT, (1960)

Krishna, A., Shenoy, R., Jha, B., Liu, Z., Paul, D., Ershaghi, I.: Repurposing idle oil and gas wells for large-scale subsurface energy storage in saline aquifers. In: SPE 209260 Presented at the 2022 SPE Western Regional Meeting, Bakersfield CA. D011S005R005 (2022). https://doi.org/10.2118/209260-MS

Miall, A.D.: Facies models. In: Stratigraphy: A Modern Synthesis. Springer Textbooks in Earth Sciences, Geography and Environment. Springer, Cham (2022). https://doi.org/10.1007/978-3-030-87536-7_4

Miller, C.C., Dyes, A.B., Hutchinson: Estimation of permeability and bottomhole pressure from bottomhole pressure buildup characteristics. JPT 91–104 (1950)

Millikan, C.V., Sidwel, V.: Bottom-hole pressures in oil wells. Trans. **92,** 194–205 (1931). https://doi-org.libproxy1.usc.edu/https://doi.org/10.2118/931194-G

Muskat, M.: Use of data oil the build-up of bottom-hole pressures. Trans. **123**, 44–48 (1937). https://doi-org.libproxy2.usc.edu/https://doi.org/10.2118/937044-G

Rzasa, M.J., Katz, D.L.: Calculation of static pressure gradients in gas well transactions of AIME. **160**(01), 100–113. https://doi.org/10.2118/945100-G.-1945

Chapter 3
Qualitative Aspects: Pattern Recognition

Introduction of the derivative function analysis (Bourdet et al. 1984), and log–log diagnostic plots have become the primary tool for well-test analysis. The main advantage of this type of data processing is the ability to reveal different flow regimes captured during the test. Flow regime analysis applies for both a drawdown and a buildup pressure transient data. Certain patterns can be developed when pressure transient data are plotted graphically on different graphic scales. We will review some expected patterns that can be recognized. In this chapter we focus on some simple system of reservoir setting. The issue of reservoir heterogeneity and its detection has been addressed by several authors. For reference purposes, we can list Ambastha et al. (1989) who discussed the pattern of partially communicating fault in composite reservoirs, Abbaszadeh, and Cinco-Ley (1995) who presented the expected pattern for a finite conductivity fault.

3.1 Wellbore-Dominated Flow

Wellbore storage is caused by the after flow of fluids into the wellbore when the well is shut-in at the wellhead. In the early flow time, reservoir responses can be usually masked or distorted by wellbore storage. The duration of this condition is dependent on the wellbore volume, and the fluid compressibility. It may be defined as:

$$C = V_{bw}^{*} C_{wb} \tag{3.1}$$

where V_{bw} is the volume of the wellbore in and C_{wb} is the compressibility of the wellbore fluid.

When the pressure readings on a gauge at the top of the well represent fluid movements in the tubing independent of formation effects, the flow is controlled by wellbore storage. As such, wellbore-dominated flow refers to a period when recorded

I. Ershaghi, *Solved Problems in Well Testing*,
https://doi.org/10.1007/978-3-031-47299-2_3

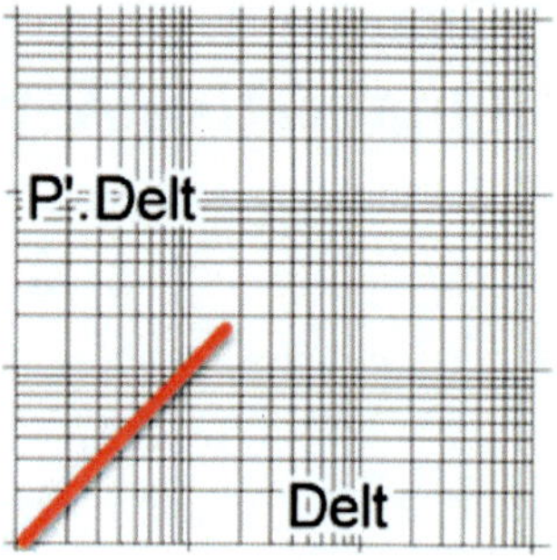

Fig. 3.1 Behaviorofconstantwell-boredominated flow on a long-log scale graph of pressure change versus time

pressure responses reflect changes inside the wellbore independent of formation contribution. The pressure vs. time data provides no clues to the nature of reservoir rock properties. But what can enhance this type of flow and its duration is the fluid type and the presence of formation skin. Higher compressibility fluids and higher skins prolong the test duration of the wellbore storage. During the wellbore-dominated flow, the formation is not controlling the flow, and as such, the pressure changes in the tubing are caused by fluid expansion and phase separation. For simple changes, the recorded pressure and time are proportional. That leads to estimation of a parameter that is called wellbore storage factor. If this factor changes with time, there is a variable wellbore storage condition. A long-log scale plot of P.' dt versus dt can be used to recognize the flow duration affected by wellbore storage. Figure 3.1 shows a typical response of wellbore storage-affected pressure response.

3.2 Manifestation of Radial Flow

A202: When flow around a wellbore is uniform from all directions, we refer to this condition as radial flow. In general, such a flow may be recognized from a plot of recorded pressure versus time on a semi-log paper or a derivative function plot on a log–log paper, Fig. 3.2.

During the process of pattern recognition, deviations from such a straight line may be indicative of conditions that represent the non-radial flow. For example, if there are high conductivity fractures imbedded in the rock, the flow is primarily through the fractures and as such, deviations from radial flow are noted particularly on a pressure derivative log–log plot, Fig. 3.3.

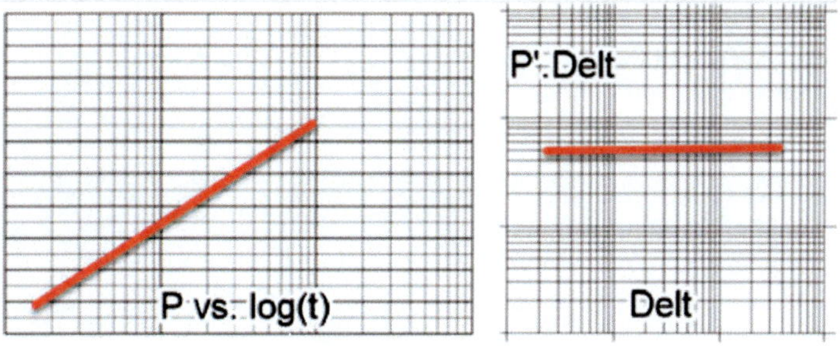

Fig. 3.2 Behavior of radial flow condition on a semi-log scale and the pressure derivative function plot on a log–log scale

Fig. 3.3 Effect of natural fractures on the pressure derivative function plot on a log–log scale

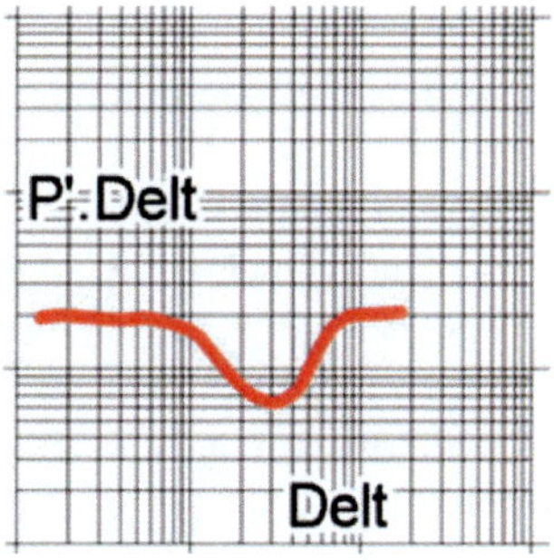

3.3 Variations in Reservoir Transmissibility

Pressure transients recorded by a downhole gauge are affected by the changes in the lateral transmissibility of formation. Both the permeability and the formation thickness affect the measured transmissibility. Changes caused by lateral variations in reservoir permeabilities, such as formation thickness or fluid viscosities, affect lateral transmissivities and may be detected by slope changes observed on the log–log pressure derivative function plot versus time.

3.4 Questions and Answers

Q3.1: Given the following response (Fig. 3.4) from a pressure buildup case. Is there any evidence that the transient is approaching the well in a radial fashion?

A3.1-Yes, a flat portion seen on the log–log pressure derivative plot is a potential indication of radial flow.

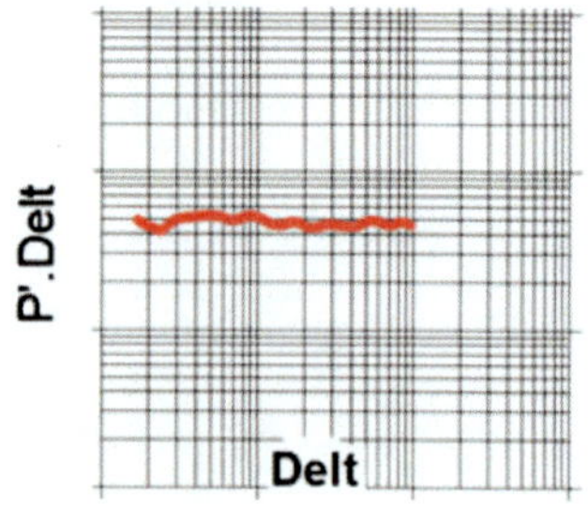

Fig. 3.4 Behavior of radial flow on a log–log plot of the pressure derivative function plot

3.5 Presence of Near Wellbore Damage

The presence of an interval with low permeability around a well can cause a restriction to flow, and the manifestation is a delay in the pressure rise during a buildup test and an increase in pressure drop during the drawdown test. On the log–log pressure derivative function plots, this is seen with a rapid change in the slope.

The following Fig. 3.5 represents the pressure derivative function plot for a buildup case. Is there any evidence that the well is damaged?

The flat portion seen indicates radial flow. The early part shows the effect of wellbore storage and skin Fig. 3.6.

The lateral variation in reservoir transmissibility can be detected on the log–log of pressure derivative function plots. A drop or an increase in transmissivities correspond to an inverse response in the pressure derivative function plots.

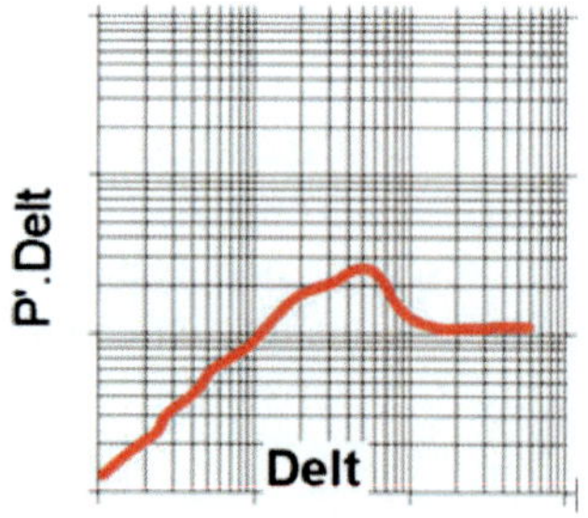

Fig. 3.5 Manifestation of wellbore storage and skin before radial flow

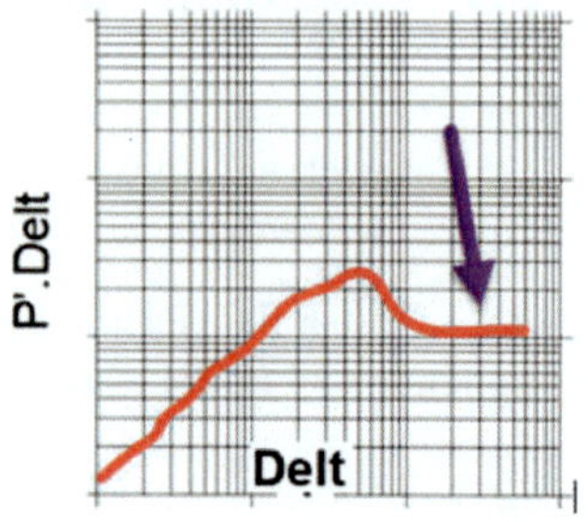

Fig. 3.6 Radial flow is seen after a period dominated by wellbore storage and skin. Lateral variations in reservoir permeability, thickness, or fluid properties

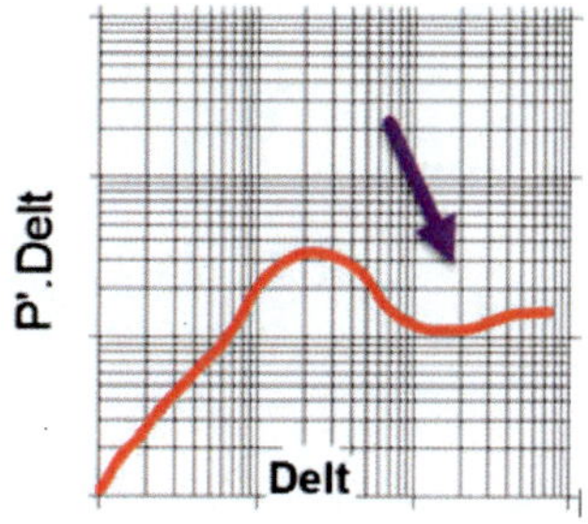

Fig. 3.7 Radial flow followed by a period of slope increase typically caused by either lateral variation of transmissibility or a no cross flow layered system

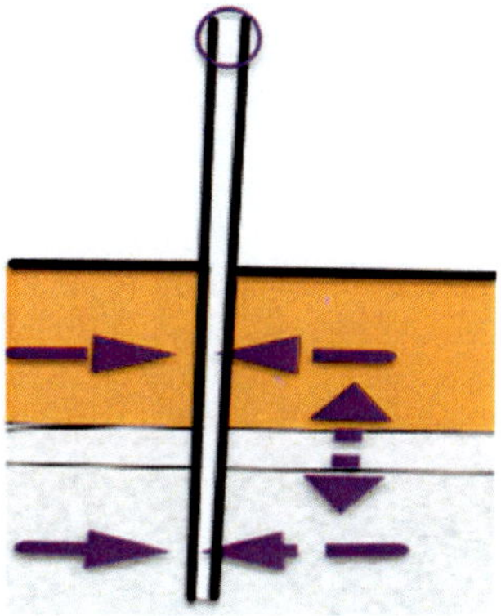

Fig. 3.8 Manifestation of crossflow in layered systems

From the following pressure derivative function plot, Fig. 3.7, for a buildup case, we can see indication of lateral changes in reservoir transmissibility.

The increase in the value of the slope indicates a lowering of transmissibility. This can be caused by lateral variation of transmissibility or presence of layered systems with no cross flows among layers or a composite system, Fig. 3.8.

3.6 Composite Reservoirs

There are cases where the area around a well may consist of varying-permeability rocks at distances from the wellbore, Fig. 3.9. A good example is the case of channel sands, like the one shown below. If a pressure transient test is conducted, one may observe a high permeability zone changing into a low permeability zone, B. These situations are best represented by composite reservoir models. Several mathematical derivations have been proposed for these conditions. Among those are formulations offered by Carter (1966). The expected pressure response can show variations on the slope of the Horner plot and on the pressure derivative function plot.

A typical Horner plot for a composite system and the corresponding pressure derivative function plot for a composite reservoir is shown in Fig. 3.10.

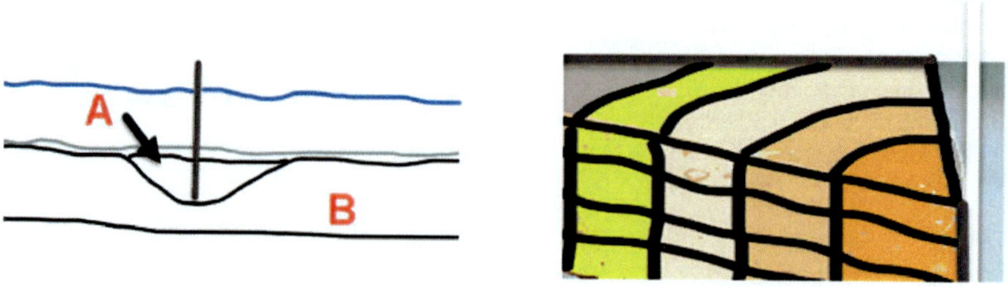

Fig. 3.9 Manifestation of a composite reservoir system

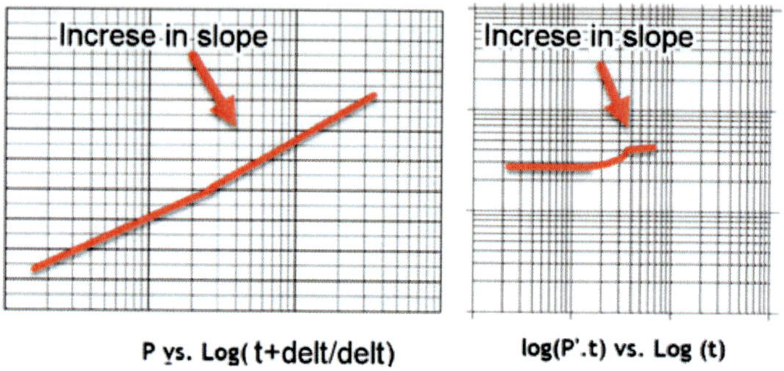

Fig. 3.10 Manifestation of slope increase on a horner plot and on a pressure derivative function plot for a composite system

3.7 The Boundary Effects

When there is a limited or no flow boundary, it may be detected on the measured pressure transients. Boundaries can prevent further transmission of pressure transients. Physical or structural changes could cause this. Signals recorded by pressure gauges reflect the pressure transient movement inside formations. If there is a no-flow boundary ahead, the transient will slow down or stop diffusing, Fig. 3.11.

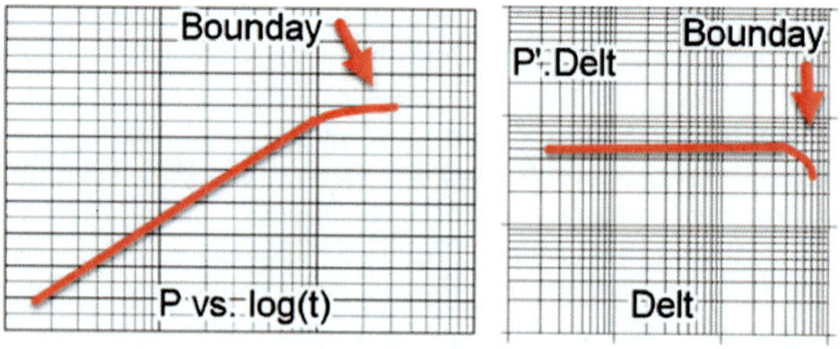

Fig. 3.11 Plots indicating boundary effects

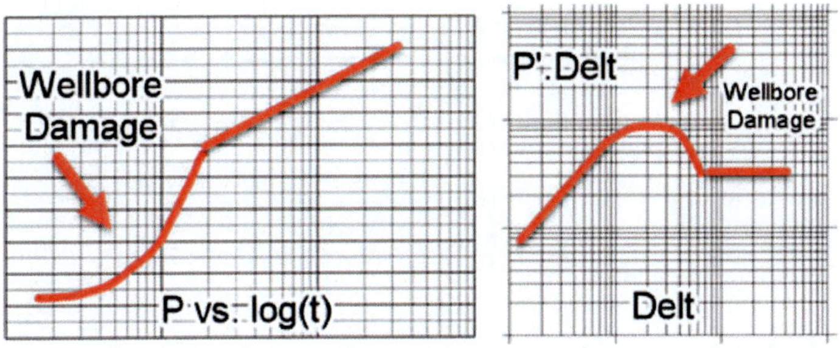

Fig. 3.12 Plots indicating infinite acting radial with wellbore storage and damage

3.8 Detection of Wellbore Damage

Wellbore damage is detected when there is a delay in formation response, this can be manifested in the recordings of pressure build-up and in the recording of pressure drawdown tests. The presence of near well bore damage results in additional pressure drops and delays in building up pressure during build up tests and in delays in establishing a true radial behavior during pressure drawdown test, Fig. 3.12.

3.9 Pattern Recognition

Figures 3.13 and 3.14 show the expected patterns on a semi-log and a log–log plot for some reservoir conditions.

Homogeneous Infinite Acting Radial Flow

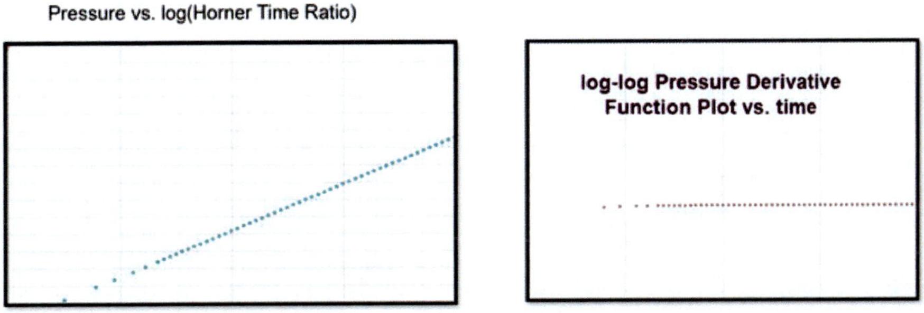

Fig. 3.13 Effect of early time without wellbore storage and skin

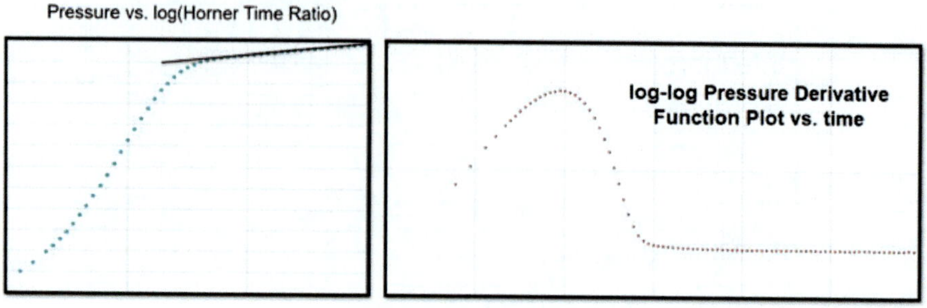

Fig. 3.14 Effect of early time with wellbore storage and skin

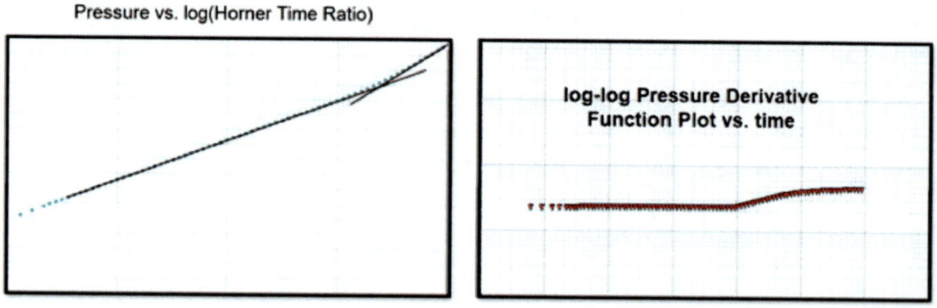

Fig. 3.15 Manifestation of a no-flow linear boundary

3.10 A Single Non-Conductive Fault

A single fault of no hydraulic conductivity manifests itself distinctly on a semi-log plot or on a derivative function plot, Fig. 3.15.

The same effects can be seen in the presence of wellbore storage and skin, Fig. 3.16.

Yaxley (1987) and Abbaszadeh and Cinco-Ley (1995) developed analytical solutions to describe transient-pressure behavior of an active well near finite-conductivity fault. They developed type curves for estimating fault skin factor and faut conductivity.

3.11 Perpendicular Faults

Perpendicular faults generate a response indicating 4-time slope changes as seen on the semi-log of pressure vs. time or on a log–log plot of a derivative function plot, Fig. 3.17.

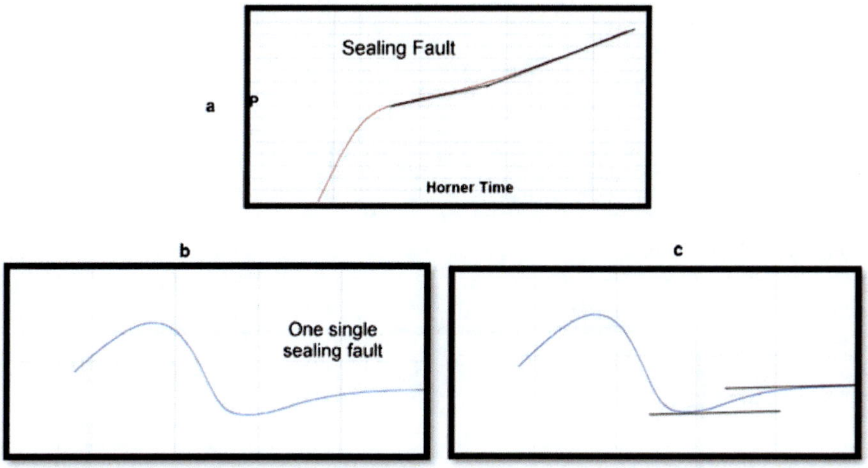

Fig. 3.16 Manifestation of a no-flow linear boundary with wellbore storage and skin. **a** Horner plot, **b** Pressure derivative function plot **c** Doubling of pressure derivative function value seen on log–log pressure derivative plot

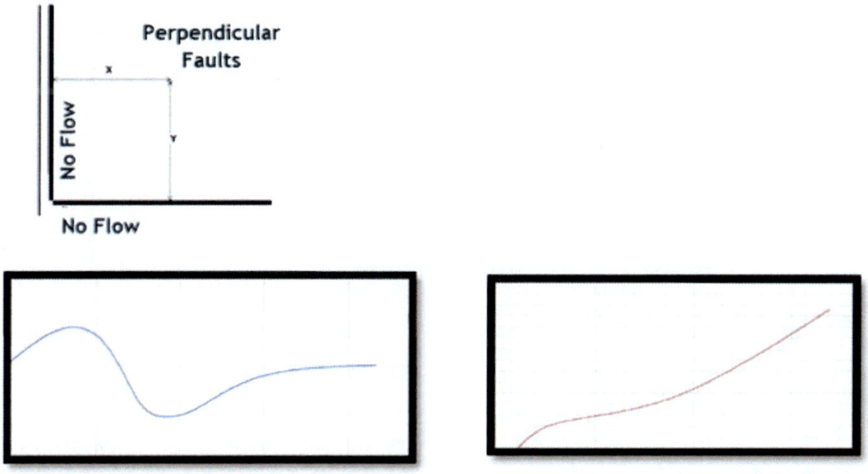

Fig. 3.17 Manifestation of two perpendicular linear boundaries with wellbore storage and skin

3.12 Parallel Faults

Parallel faults show for some time a redial flow followed by a linear flow, and these effects can be seen on a semi-log plot and the log–log plot of the pressure derivative function plot, Fig. 3.18.

Below we also show the response effect of wellbore storage, and skin included at the start on a Horner and pressure derivative function plot, Fig. 3.19.

Parallel
Faults

Fig. 3.18 Parallel faults

a

b

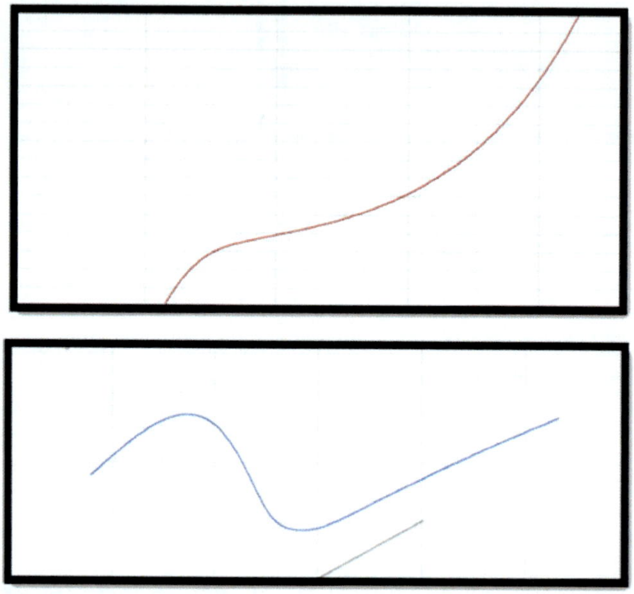

Fig. 3.19 Pressure transient response of parallel faults, a Semil log plot, b Log–Log pressure derivative plot

If the two parallel faults have a wide spacing, the above effects may not be seen in short-duration tests, and only the initial radial flow is exhibited, Fig. 3.20.

Ambastha et al. (1989) discussed the hydraulic characterization of faults. They modeled such conductivity by using an infinitesimal-thickness skin boundary. In the presence of faults, we observe pressure build-up tests can identify discontinuities that exist in subsurface geologic structures. When these are encountered in reservoirs of fluids, there are concerns about leakage across these discontinuities. Pressure build-up tests can recognize no flow faults.

The Log–Log pressure derivative function plot for a semi-sealing fault is shown in Fig. 3.21.

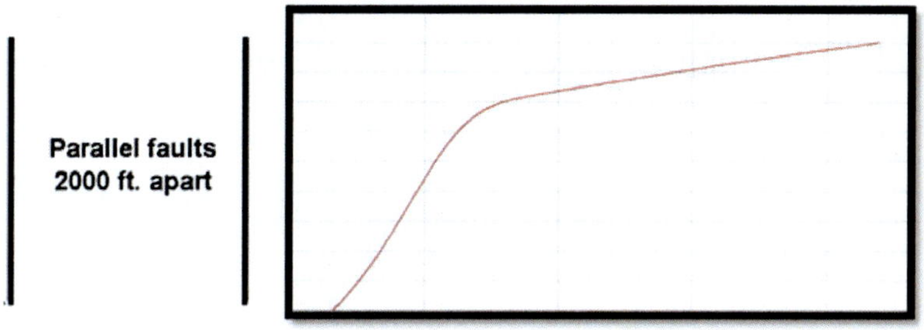

Fig. 3.20 Horner plot response with the wellbore storage and skin with large spacing between the two parallel faults

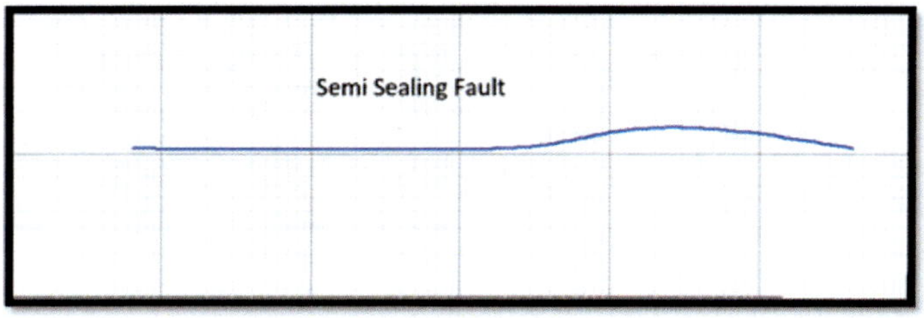

Fig. 3.21 Log–Log plot of the derivative function plot for a semi-sealing fault

3.13 Boundary Effects

When boundaries are detected, there is a noticeable change in the pressure derivative function plot. A close boundary results in a significant drop in the log–log vs. time of a pressure derivative function plot. For boundaries which are associated with oil–water contacts or gas-oil contacts, there is a change in the slope of the pressure derivatives function plot. As other conditions, such as lateral variation, also cause a similar change, one needs to use various sources of information about the fluid boundary changes.

Depending on the duration of the test, one or more boundaries may be detected. If the boundaries are of a no-flow type, such as a sealing fault, the response has features indicating approaching a zone or reduced permeability. If the boundaries reflect aquifer or gas cap support, the change in the fluid mobility can be sensed. The effect of a closed boundary can be seen both on the semi-log plot and the log–log derivative function plot, Fig. 3.22.

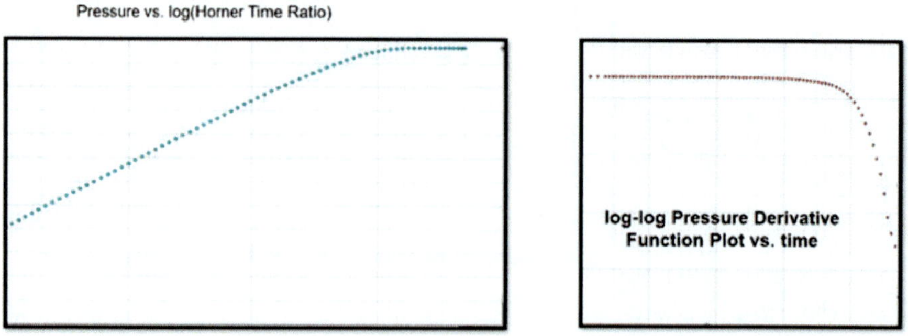

Fig. 3.22 Semi-log plot and the log–log derive function plot of a bounded reservoir

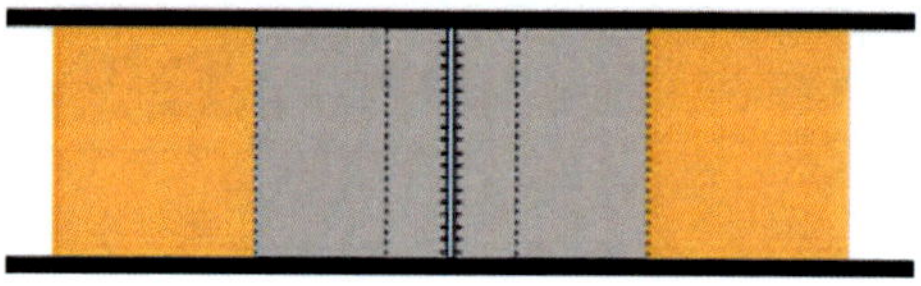

Fig. 3.23 Variable properties in a composite reservoir

3.14 Lateral Variations

The lateral variation in reservoir transmissivities is manifested by a composite reservoir effect, as shown in Fig. 3.23.

Figure 3.24 shows the responses on a Horner and a pressure derivative function plot.

3.15 Layered Reservoirs

When testing layered reservoirs, the important thing to note is the existence or nonexistent crossflow among the layers. With crossflow, a series of layers may respond equivalently to a single layer, Fig. 3.25. if there are no crossflows, the entire system may respond similarly to a single layer. As indicated by Ehlig- Economides and Joseph (1985) only if flow rates from individual layers are measured one can find a way to examine the properties of individual layers.

Log–Log pressure derivative function plot can be diagnostic for a composite reservoir with increasing permeability from the well (Fig. 3.26).

For example, the Log–Log derivative function plot looks for a composite reservoir with decreasing permeability from the well, may show a pattern like Fig. 3.27.

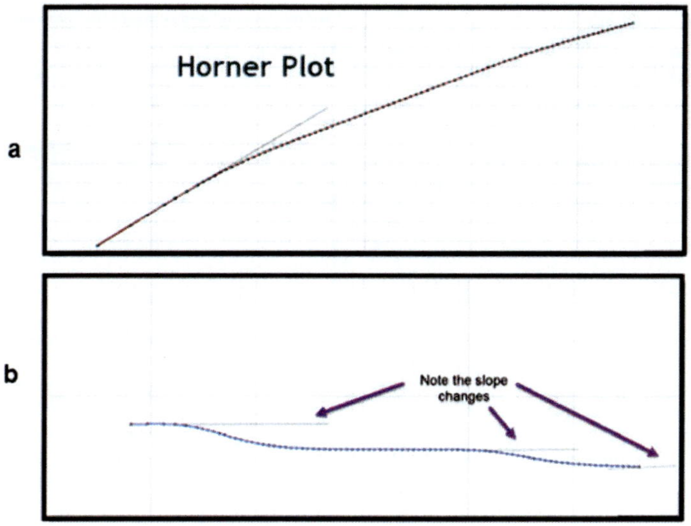

Fig. 3.24 **a** Semil log plot of P versus horner time and **b** Log–Log plot of pressure derivative function behavior of a composite reservoir

Fig. 3.25 **a** Horner and **b** Derivative function plots for typical layered system

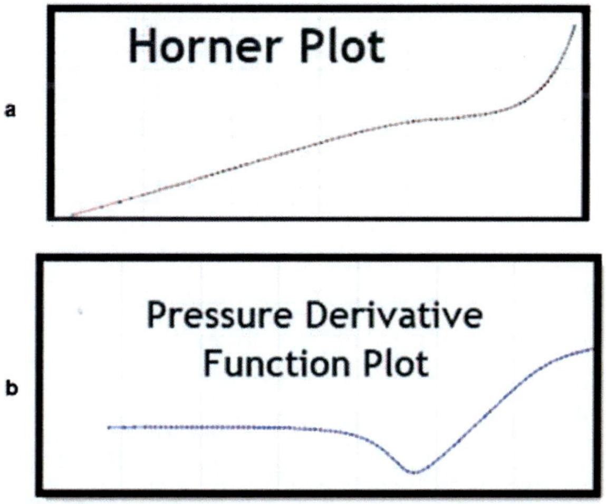

3.16 Well Test Results for Slanted Wells

Pressure buildup test data for slanted holes can exhibit negative skins. Such pseudo skins may mask the true skin that may exist around the wellbore. In slant a negative skin factor. The mathematical reasoning for this was developed by Cinco-ley et al. (1975).

a

b

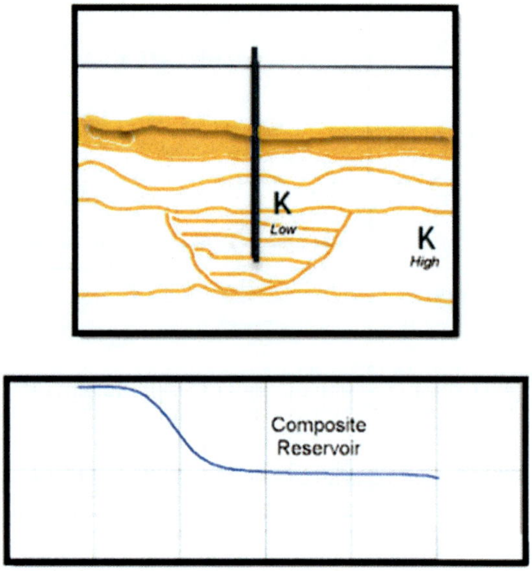

Fig. 3.26 **a** Reseervoir image and **b** Corresponding pressure derivative function plot behavior of composite system from low transmissibility to high transmissibility

a

b

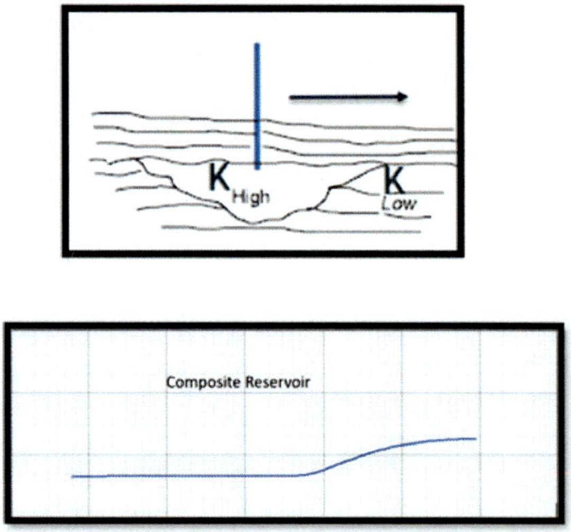

Fig. 3.27 **a** Reservoir image and **b** Corresponding pressure derivative function plot behavior of composite system from high transmissibility to low transmissibility

3.17 Other Cases

Other cases are discussed in the Chapters ahead. These include test data for naturally fractured reservoirs, that when well tests are conducted with adequate duration, we may detect the fracture systems and the transition that reflects the recharging of fractures by the matrix system. Other interpretation solutions have been published indicating that systems with various degrees of fracturing can be detected from well-tested responses.

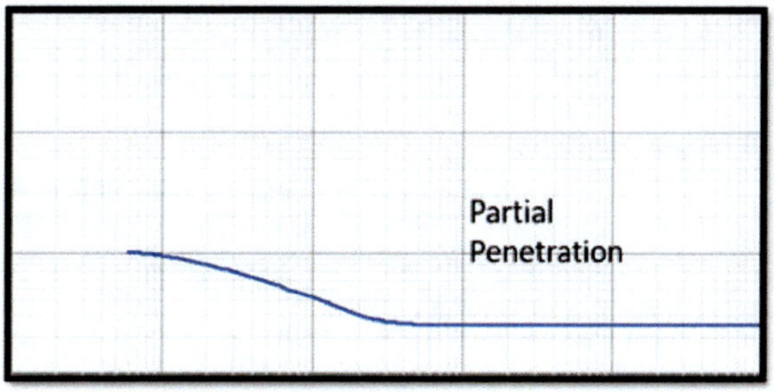

Fig. 3.28 Behavior of partially penetrating well

Review of test data for horizontal wells is covered Chap. 6 and the test data can indicate distinct segments. Initially, a vertical radial flow is seen, followed by a lateral related linear flow and eventually a formation pseudo radial flow.

Pressure build-up tests on well that have been hydraulically fractured show the dominance of fracture flow. We will also discuss hydraulically fractured wells and show, based on predictions published by Agarwal et al. (1979) that the slopes of the derivative plots approached ½ indicating high conductivity flow and those with slopes less than one-half representing fractures of limited conductivity.

We will also discuss drill stem tests which are valuable measurements that can provide information about damages caused by drilling and help to estimate reservoir pressure and sometimes reservoir configurations.

3.18 Questions and Answers

Q3.2: How would the Log–Log pressure derivative function plot look for a partially penetrating well?

A3.2: See Fig. 3.28.

Q3.3: What is the difference between a Horner plot and MDH plot >

A3.3: For long producing time we can use P versus log(dt), and this would generate the same slope as plotting P versus log $(\frac{tp+dt}{dt}\}$ as used in a Horner plot.

References

Abbaszadeh, M., Cinco-Ley, H.: Pressure-transient behavior in a reservoir with a finite-conductivity fault. SPE Form Eval. **10**, 26–33 (1995). https://doi-org.libproxy1.usc.edu/https://doi.org/10.2118/24704-PA

Agarwal, R.G., Carter, R.D., Pollock C.B.: Evaluation and performance prediction of low-permeability gas wells stimulated by massive hydraulic fracturing. J. Pet Technol. **31**(03), 362–373 (1979). Paper Number: SPE-6838-PA https://doi-org.libproxy1.usc.edu/https://doi.org/10.2118/6838-PA

Ambastha, A.K., McLeroy, P.G., Grader, A.S.: Effects of a partially communicating fault in a composite reservoir on transient pressure testing. SPE Form Eval. **4**, 210–218 (1989). https://doi-org.libproxy1.usc.edu/https://doi.org/10.2118/16764-PA

Bourdet, D., Ayoub, J.A., Pirard, M.: Use of pressure derivative in well-test interpretation. SPE Form Eval. **4**(02), 293–303 (1984) Paper Number: SPE-12777-PA https://doi-org.libproxy1.usc.edu/https://doi.org/10.2118/12777-PA

Carter, R.D.: Pressure behavior of a limited circular composite reservoir. SPE J. **6**(04), 328–334 (1966). Paper Number: SPE-1621-PA 1966 https://doi-org.libproxy1.usc.edu/https://doi.org/10.2118/1621-PA

Cinco-ley. H., Ramey, H.J., Miller F.G.: Pseudo-skin factors for partially penetrating directionally drilled wells. In: Paper presented at the Fall Meeting of the Society of Petroleum Engineers of AIME, Dallas, Texas, September 1975. Paper Number: SPE-5589-MS (1975). https://doi-org.libproxy1.usc.edu/https://doi.org/10.2118/5589-MS Published: September 28, 1975

Ehlig-Economides, C.A., Joseph, J.: A new test for determination of individual layer properties in a multilayered reservoir. SPE Form. Eval. **2**(03), 261–283 (1985). https://doi.org/10.2118/14167-PA,1985

Horner, D.R.: Pressure buildup in wells. In: Third World Petroleum Congress. Leiden, The Netherland, Section II, Preprint 7, pp. 25–43. (1951)

Miller, C.C., Dyes, A.B., Hutchinson: Estimation of permeability and bottomhole pressure from bottomhole pressure buildup characteristics. JPT 91–104 (1950)

Yaxley, L.M.: Effect of a partially communicating fault on transient pressure behavior. In: SPE Formation Evaluation Paper Number: SPE-14311-PA SPE Form Eval vol 2(04). pp. 590–598. (1987). https://doi.org/10.2118/14311-PA

Chapter 4
Quantitative Methods

In this Chapter, we summarize some of the equations and relationships that can assist in estimating various properties from well test data. The following three fundamental equations represent the relationship between the measures pressure and reservoir properties in pressure drawdown, pressure build up and interference tests.

4.1 Pressure Buildup and Pressure Drawdown

Using the English set of units, the four main equations that can be used in parameter estimation relating to radial flow system a can be written as:

Wellbore flowing pressure:

$$p_{wf} = pi - 162.6\left(\frac{\mathrm{q} \cdot \mathrm{B} \cdot \mu}{\mathrm{k} \cdot \mathrm{h}}\right)\left(\log \mathrm{t} + \log(\frac{k}{\Phi \cdot \mu \cdot ct \cdot rw^2}) + 0.8686\, S - 3.2274\right) \tag{4.1}$$

Wellbore shut-in pressure

$$p_{ws}(\Delta t) = pi - 162.6\left(\frac{\mathrm{q} \cdot \mathrm{B} \cdot \mu}{\mathrm{k.h}}\right)\log\left(\frac{\mathrm{tp} + \Delta \mathrm{t}}{\Delta \mathrm{t}}\right) \tag{4.2}$$

Skin factor:

$$\mathrm{sk} = 1.151\left(\frac{\mathrm{pi{-}p1hr}}{\mathrm{m}}\right) - \log\left(\frac{k}{\Phi \cdot \mu \cdot ct \cdot rw^2}\right) + 3.2274 \tag{4.3}$$

$$\Delta p = \frac{-70.6\ q.\ \mu.B}{k.\cdot h}\mathrm{Ei}\left(\frac{-948\ \emptyset.\mu.c.\ r^2}{k \cdot t}\right) \tag{4.4}$$

Equation (4.4) is also used for interference pressure effect calculations.

I. Ershaghi, *Solved Problems in Well Testing*,
https://doi.org/10.1007/978-3-031-47299-2_4

Assuming the dominance of radial flow, Eqs. 4.1 and 4.2 show the relationships between the measured pressure and time for pressure drawdown and pressure buildup tests, respectively.

Equating 4.2 is the format of the pressure buildup equation first presented by Horner. Equation 4.3 provides estimation of skin factor and Eq. 4.4 allows estimation of pressure drop at an observation well located at a distance of (r) from the well. These equations may be written in terms of (T) Transmissibility and (S), Storativity.

4.2 Wellbore Flowing Pressure

$$p_{wf} = pi - 162.6\left(\frac{\mathrm{q.B.}}{\mathrm{T}}\right)\left(\log \mathrm{t} + \log\left(\frac{T}{S \cdot \mu \cdot rw^2}\right) + 0.8686\,\mathrm{s} - 3.2274\right) \tag{4.5}$$

Wellbore shut-in pressure

$$p_w(\Delta t) = pi - 162.6\left(\frac{\mathrm{q.B}}{\mathrm{T}}\right)\log\left(\frac{\mathrm{tp} + \Delta\mathrm{t}}{\Delta\mathrm{t}}\right) \tag{4.6}$$

Skin factor:

$$\left(sk = 1.151\left(\frac{\mathrm{pi} - \mathrm{p1hr}}{\mathrm{m}}\right) - \log\left(\frac{T}{S \cdot \mu \cdot rw^2}\right) + 3.2274\right) \tag{4.7}$$

$$\Delta p = \frac{-70.6\ q\ B}{T}\mathrm{Ei}\left(\frac{-948\ S\ r^2}{T \cdot t}\right) \tag{4.8}$$

How is the average reservoir pressure computed?

A-When a well is shut in for a long time, a stabilized pressure may be estimated from a buildup test. That estimation represents the average pressure in the drainage area of the well. To get the reservoir average pressure, assuming perfect connectivity, all wells must be shut in and the average reservoir pressure can represent a weighted average of average pressure for individual wells. For individual wells, if the test duration is not long, a method such as that proposed by (MDH) Mathews et al. (1954), can help in the estimation of average pressure.

$$P\ (\mathrm{average}) = \frac{p1 \cdot q + p2 \cdot q2 + P3 \cdot q3}{q1 + q2 + q3+} \tag{4.9}$$

4.3 Use of Fiber Optics

One of the latest technologies in monitoring reservoir pressures is the use of fiber-optics. This involves the use of optical fibers as sensors to detect and monitor changes in pressure, temperature, and other parameters. It enables real-time, continuous monitoring of reservoir pressures and other key parameters. Fiber-optic sensing has been used to cases where signal has not been detectable by conventional monitoring techniques.

It includes a long, continuous fiber-optic cable that is installed in the wellbore or surrounding formation, with associated hardware and software for data acquisition and analysis. The optical fibers in the cable act as distributed sensors that can detect changes in temperature, strain, and pressure, and transmit this data to the surface for real-time monitoring and analysis.

It can provide a promising approach for monitoring pressures in subsurface carbon dioxide injection, allowing optimized injection practices and improving storage efficiency.

4.4 Reservoir Permeability Computations

A-In all the pressure transient analysis methods, what is estimated is transmissibility.

(T) that includes the thickness and fluid viscosity besides the effective permeability.

$$T = \frac{k \cdot h}{\mu} \tag{4.10}$$

From the computation of transmissibility, one can obtain (k) if estimates of (h) and (μ) can be provided.

4.5 Questions and Answers

Q4.1: How is the formation damage computed?

A4.1: Formation damage is expressed as the skin factor. The skin factor can best be estimated from pressure build-up tests. Besides formation damage, other causes may contribute to the estimated skin factor. These include pseudo skin caused by perforation plugging, partial completion, angle of the hole and flow turbulence, mostly in gas wells. It is important that these effects are taken out to estimate the true skin factor indicating formation damage.

Q4.2: How do we compute the skin factor from pleasure buildup tests?

A4.2: The total skin factor can preferably be estimated from the analysis of pressure build data from a Horner plot.

$$sk = 1.151\left(\frac{\text{pi}-\text{p1hr}}{\text{m}}\right) - \log\left(\frac{T}{S \cdot \mu \cdot rw^2}\right) + 3.2274) \quad S = \varphi.ct.h \tag{4.11}$$

Q4.3: How can we compute the distance to a no-flow boundary?

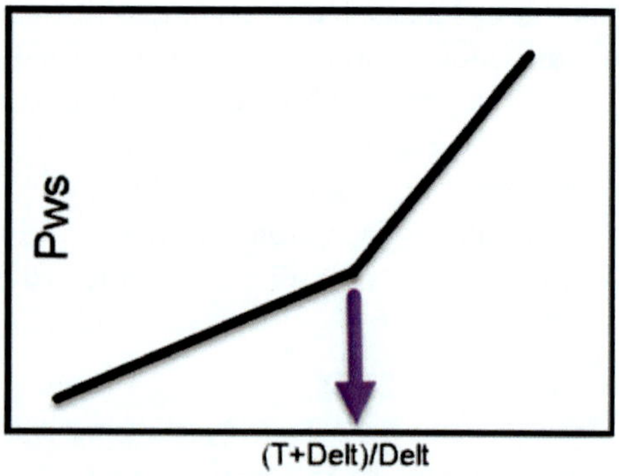

Fig. 4.1 The time corresponding to the intersection of the two straight lines

A4.3: From a pressure buildup test, the presence of a no flow boundary can be detected with a changing slope and the time corresponding with that (Delt) that can be used to estimate the distance from similar formulas for the radius of investigation, Fig. 4.1.

$$d = \sqrt{\frac{0.000148\, k.\Delta t_x}{\varphi.\mu.Ct}} \tag{4.12}$$

where Δt_x is the intersection time of the two straight lines, Gray (1965).

Q4.4: How can we estimate the wellbore storage factor?

A4.4: If there is evidence of wellbore storage, the early time data can help in estimating the wellbore storage factor. On a log–Log plot of pressure changes with time, the early time data falling on a unit slope can be considered points affected by the wellbore storage. The following formula can help in estimating the wellbore storage factor, Fig. 4.2.

$$C = \frac{q \cdot B \cdot dt}{24 \cdot dp} \tag{4.13}$$

Q4.5: How can we estimate the size of the drainage area?

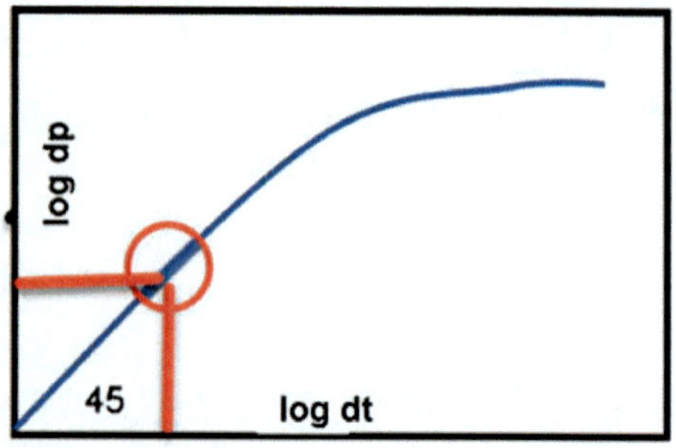

Fig. 4.2 Estimation of wellbore storage constant from a point on the line with a slope of 45°

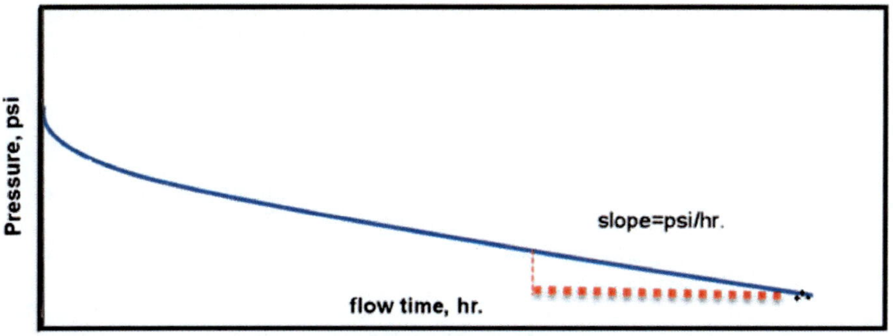

Fig. 4.3 Estimation of drainage volume from the slope of the semi steady state period of a pressure drawdown data

A4.5: The drainage area can best be estimated from a pressure drawdown test. If the test duration is long, a plot of pressure versus time on a cartesian scale may develop a constant slope. If that is observed, it can point to a semi-steady state condition and the slope of the line in psi/hr. can be used in the following formula to estimate the drainage volume and from that drainage area can be estimated (Fig. 4.3).

$$\text{Drainage volume in Barrels} = \frac{\text{q.B}}{24^{*}\text{Ct.(slope)}} \tag{4.14}$$

Q4.6: How do we estimate the equivalent dimensions of a hydraulic fracture?

A4.6: We may use an Agarwal et al. type-curve and after matching the well test data, we can estimate the Xe that represents the length of a hydraulic fracture.

Answer: The line source solution is:

$$\text{Pd}(\text{r}_\text{d}, \text{td}) = -\frac{1}{2}\,\text{Ei}\left(-\frac{r_d^2}{4.t_d}\right)$$

where −Ei((-x) can be obtained from a plot as shown below/

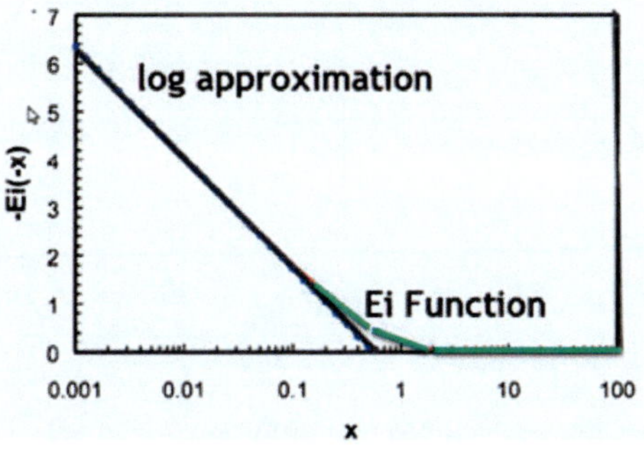

$$\text{For } x < 0.01 \quad -\text{Ei}(-x) = -\ln(1.78 \cdot X)$$

$$\Delta p = -70.6\frac{q.\mu.B}{k.h}E_i\left(-\frac{948\varphi.\mu.C_t.r^2}{k.t}\right) \tag{4.15}$$

Q4.7: How do we estimate the transmissibility and storativity between two wells undergoing an interference test?

The graphical solution is shown in Fig. 4.4.

A4.7: For the log–log plot using the dimensionless type curve, by matching the actual data points, we can estimate a match point on the Pd versus td plot, which is the relationship of Pd to td for an infinite acting reservoir that can lead to the estimated transmissivity and storativity.

$$\text{Pd}(\text{r}_\text{d}, \text{t}_\text{d}) = -\frac{1}{2}\,\text{Ei}\left(-\frac{r_d^2}{4.t_d}\right) \tag{4.16}$$

Here is an example on an actual test in the Anschutz Ranch East Field from Pollock and Bennet (1986) (Fig. 4.5).

Q4.8: How do we estimate the direction of highest transmissibility in an anisotropic reservoir?

A4.8: There needs to be at least four wells geometrically covering various directions surrounding the pulsing well. Interference or pulse tests between the pulsing wells and other wells can identify the presence of directional change in the transmissibility.

The formulation was presented by Ramey (1975) and further discussed by Pan et al. (2018) (Fig. 4.6).

$$K_{\text{effective}} = \sqrt{k_{xx} \cdot k_{yy} - k_{xy}^2} \quad \text{(a)}$$

$$\text{Pd} = \frac{h(P_i - p_{x,y,t})}{141.2\,q\mu B}\sqrt{k_{xx}, k_{yy} - k_{xy}^2} \quad \text{(b)}$$

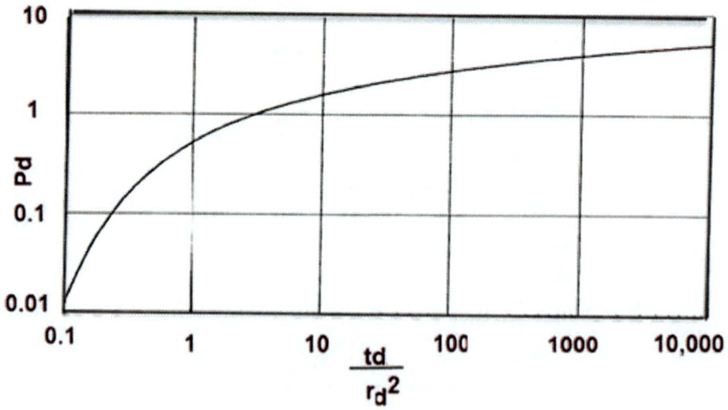

Fig. 4.4 Graphical solution of the line source diffusivity equation in a radial system

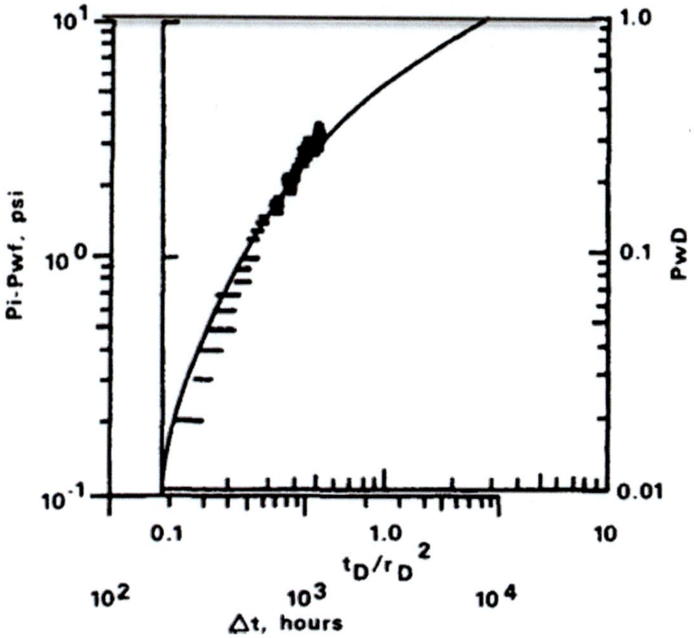

Fig. 4.5 Type curve matching of actual data versus the line source solution. (Courtesy SPE -Pollock and Bennet 1986)

Fig. 4.6 Representation of anisotropy in permeability

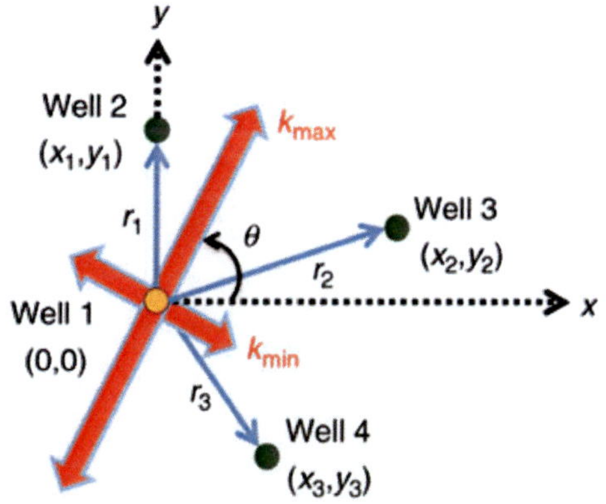

$$\text{K effective} = \sqrt{k_{xx}, k_{yy} - k_{xy}^2} \quad \text{(c)}$$

$$\frac{td}{r_d^2} = \left(\frac{0.000264 \cdot t}{\emptyset\mu C_t}\right)\left(\frac{k_{xx} \cdot k_{yy} - k_{xy}^2}{Y^2 k_{xx} - 2x \cdot y \cdot k_{xy} + X^2 k_{yy}}\right) \quad \text{(d)}$$

$$\text{K}_{\text{max}} = \left[\frac{1}{2}\left[(k_{xx} + k_{yy})\right] + \left[\sqrt{(k_{xx} - k_{yy})^2 + 4k_{xy}^2}\right]\right] \quad \text{(e)}$$

$$\text{K}_{\text{min}} = \left[\frac{1}{2}\left[(k_{xx} + k_{yy})\right] - \left[\sqrt{(k_{xx} - k_{yy})^2 + 4k_{xy}^2}\right]\right] \quad \text{(f)}$$

$$\Theta = \arctan\left(\frac{kmax - k_{min}}{k_{xy}}\right) \quad \text{(g)} \tag{4.17a-g}$$

Q4.9: What is the pressure drop at a well because of production at a nearby well?

A4.9: Delp $= -\frac{70.6q.Bo}{T}$ Ei $[\frac{-948S.r2}{T.t}]$

$$-\text{Ei}(-\text{x}) = \int_{x}^{\infty} \frac{e^{-u}}{u} du \text{ for x } < 0.02 \quad -Ei(-x) = -2.303[\log(x) + 0.25]$$

Q4.10: How is the distance to a boundary computed?

A4.10: d $= 0.012[\sqrt{\frac{k.tx}{\emptyset.\mu.Ct}}]$ Use equation 4.2 where tx is the time, it takes for the boundary to be detected.

Q4.11: Does the radius of investigation depend on flow rate?

A4.11: In the equation for radius of investigation, the flow rate does not show up.

In the radius of investigation equation, r $= \sqrt{\frac{0.00105\ k \cdot tp}{\Phi.\mu.ct}}$, no production or injection rate shows up, but obviously, there has to be some flow or injection rate to measure a detectable pressure change at a distance

Q4.12: If the schedule of rate change is known, what is the formula to compute the well flowing pressure; q1, q2, q3 are rates.

A4.12: The formulation is shown below:

$$\text{Pt} = \text{Pi} - \frac{162.6q1.B}{T}\log(t) - \frac{162.6(q2 - q1).B}{T}\log(t - t1) - \frac{162.6(q3 - q2).B}{T}\log(t - t2) + \ldots \tag{4.18}$$

References

Earlougher, R.C., Kazemi, H.: Practicalities of detecting faults from buildup testing. JPT 18–20 (1980)

Gray, K.E.: Approximating well-to-fault distance from pressure build-up tests. JPT **7** 761–767 (1965)

Matthews, C.S., Brons, F., Hazebrock, P.: A method to determination of average pressure in a bounded reservoir, AIME (1954)

Pollock, C.B., Bennett, C.: Eight-well interference test in the anschutz ranch east field. SPE Formation Evaluation (1986)

Ramey, H.J.: Interference analysis for anisotropic formations—a case history (includes associated paper 6406). J. Pet Technol. **27**(10), 1290–1298 (1975). SPE-5319-PA. https://doi.org/10.2118/5319-PA

Theis, C.V.: The relation between the lowering of the piezometric surface and the rate and duration of discharge of a well using groundwater storage. Am. Geophys. Union Trans. **16**, 519–524 (1935)

Yan, P., Kamal, M., Narr, W.: Fieldwide determination of directional permeabilities using transient well testing. J.: SPE Reserv. Eval. Eng. **22**(02), 734–744 (2018). Paper Number: SPE-181437-PA 2018

Chapter 5
Gas Wells

The analysis of gas well test data includes seeking answers to several questions such as the present deliverability, the future deliverability at various reservoir and well-flowing pressures and the impact of the completion efficiency for obtaining higher deliverability. Other benefits include if there are sufficient gas reserves indicated to justify additional development wells. Also, there are needs for the estimation of realistic reserves by the pipeline company and the most accurate reserve for field development planning and fair market value evaluation, (Jones 1962).

Testing of gas wells is important to assess the reservoir characteristics, the productivity, and for evaluating the suitability for commercial production of natural gas. Usually, gas wells are equipped with wellhead facilities that control the flow of gas during testing. Before testing, the well may undergo an initial clean-up process for removing drilling mud and other debris from the wellbore. Several methods can be used to measure and monitor the flow rate of gas during testing including flow meters, pressure gauges, orifice plates, or sonic meters.

Pressure build-up test starts with shutting in the well and monitoring the pressure response for the estimation of permeability, pressure depletion, and boundary conditions. For checking the flowing conditions, the gas is allowed to flow to the surface at different rates and by measuring the gas flow rate and well flowing pressure, we can determine the well's productivity and reservoir parameters. At times, the wells can be shut in to measure pressure buildup over a period to determine reservoir deliverability and pressure support.

During testing, samples of the produced gas and liquids may be collected for analysis to determine the composition, number of impurities, and other fluid properties. The collected data can be analyzed to assess reservoir performance, estimate reserves, optimize production strategies, and make decisions regarding the commercial viability of the well.

Gas well testing procedures may vary depending on well characteristics, regional regulations, and the objectives of the specific testing program. As such, the testing methods and considerations can differ from one gas well to another. It is important

I. Ershaghi, *Solved Problems in Well Testing*,
https://doi.org/10.1007/978-3-031-47299-2_5

to note that gas well testing procedures may also vary depending on well's characteristics, regional regulations, and the objectives of the specific testing program. Therefore, the testing methods and considerations may differ from one gas well to another.

Interpreting pressure transient test for gas well is like the interpretation methods used for oil wells. The main difference is recognizing the pressure dependency of gas properties to pressure. As such, a pseudo pressure function was introduced by Al Hussainy and Ramey (1966) that allows the consideration of such dependencies.

The pseudo pressure is defined as:

$$m(P) = 2\int_{psc}^{p} \frac{P}{\mu z} dp \tag{5.1}$$

where psc is the pressure at standard condition. In this formula, P is the reservoir pressure, z and μ are the gas deviation factor and gas viscosity, respectively. If the term $2p/(\mu z)$ is linear in p2, m(p) becomes proportional to p2. As such, tests conducted and analyzed a long time ago, may have used pressure or pressure squared plots. It turns out that, for low pressure gas or steam in geothermal reservoirs, the use of pressure squared will be equivalent to using pseudo pressures. Also, for very high-pressure gas reservoirs, use of pressure data can give results like that used the pseudo pressure data.

In testing gas wells, a main goal is determining the present and future deliverability. This can also provide information for the requirements set by regulatory agencies. Periodic testing can also be used to meet the requirements of pipeline companies regarding the specification of deliverability. The gas well test can also evaluate the efficiency of completion with respect to the presence or worsening of the damaged zone around wells.

Such tests can help select production equipment to maximize revenues from gas supply lines so that delivery can be maintained during extreme weather conditions. Among these tests we can mention reservoir limit tests, back pressure tests for estimation of the absolute open flow potential and tests that help in determining gas reserves, Rawlins and Schellhardt (1936).

The first step is to determine the initial bottom hole pressure by means of a bottom hole pressure gauge while the well is closed. This procedure will determine if there are any fluids in the wellbore and will detect the presence of any fill from previous treatments. In addition to the pressure transient tests, the methodology using back pressure measurements is a part of well testing. In brief to get the maximum expected rate for a gas well, various back pressures are applied, and the flow rates are measured. From this information, one can generate the back pressure equation and estimate the absolute open flow potential.

$$q = C * (P^2 - P_{wf}^2)^2 \tag{5.2}$$

An improvement of this procedure is the use of an isochronal test. This procedure more correctly determines the back pressure curve. The modification suggested by

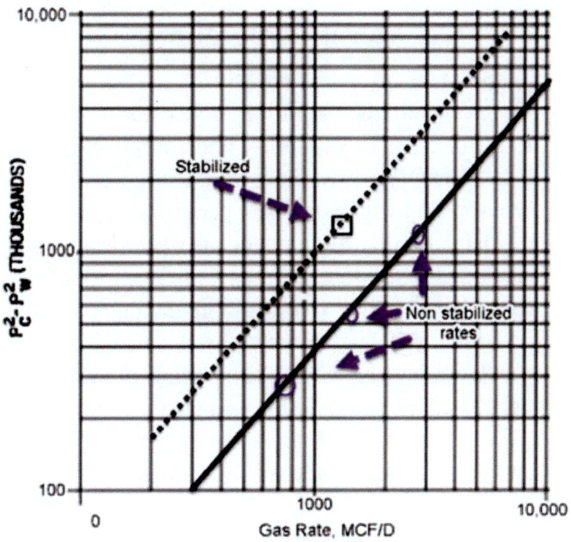

Fig. 5.1 Example for an Isochronal test plot

Cullender (1955), recognized the effect of test duration and its relation to investigating the drainage area from the back pressure data and thus led to modifications in the technology of back pressure testing. With this modification, a well is flowed for a long period to establish a point corresponding to the maximum reservoir volume influence. This is then followed by a series of short flow that can allow the derivation of back pressure equation. The standard multi-point test consists of several flow rates in ascending sequence while recording the surface and/ or bottom hole pressure data. Figure 5.1 shows the results of a four-point test obtained on a deep gas well.

Pressure drawdown tests are also used to determine the deliverability tests and formation permeability, and to estimate the minimum amount of gas reservoir connected to the well from reservoir limit tests. They can be used to also evaluate completion efficiency,

Multipoint tests are conducted for the purpose of determining the absolute open flow potential and the values of "n" for the back pressure equation. The absolute open flow potential is determined directly from the back pressure equation by equating Pwf to 14.7 psia.

$$q = C*\left(P^2 - Pwf^2\right)^n$$

5.1 Gas Storage Operations

Development and operation of gas storage fields has a long history, (Katz 1959). In terms of energy availability and energy storage, these operations play an important role in many parts of the world. Depleted or semi-depleted oil and gas fields can be repurposed for gas storage operations.

Deliverability maintenance is a primary objective of gas storage operators. Some subsurface gas storage wells exhibit a loss of deliverability over time. This can be related to an increase in wellbore damage. The damage can develop during injection or withdrawal operations or because of a change in flow direction. Examining the trends in damage versus time using conventional buildup and falloff tests can be operationally challenging and costly. One may use electronic flow measurement (EFM) systems, Dunham and Trosper (1992) for collecting high-frequency pressure and flow rate data at the wellhead when over the course of normal storage operations, when wells are shut in for short periods of time. Wellhead data during these shut-in periods can be analyzed as buildup or fall-off tests. We can filter high-frequency EFM data from gas storage wells to identify shut-in periods that can provide buildup or falloff tests and allow pressure transient analysis of the test data for generating semi-log and log–log pressure derivative function plots. Also, plots of apparent skin vs. flow rate and vs. time can help to identify flow rate sensitivities.

5.2 Questions and Answers

Q5.1: How do we estimate the absolute open flow potential with the gas well back pressure equation?

$$q = C*\left(P^2 - Pwf^2\right)^n$$

A5.1: Let $P_w f$ = atmospheric pressure, i.e., 14.7 psia.

Q5.2: How do we estimate the radius around a gas well in a gas condensate reservoir that is experiencing condensate buildup?

A5.2: A log–log plot of pressure derivative function versus (dt) may show a slope increase indicating the formation of a condensate zone at some desistance from the well.

Q5.3: What is a main cause of pressure drop sometimes observed in average pressures of gas storage fields?

A5.3: Leaks through poorly cemented injection or production wells.

References

Al-Hussainy, R., Ramey, H.J.: Application of real gas flow theory to well testing and deliverability forecasting. JPT 637-42, Trans, AIME **237** (1966)

Cullender, M.: The isochronal performance method of determining the flow characteristics of gas welis. Trans AIME **204**, 137 (1955)

Dunham, R.D., Trosper, R.L.: An integrated electronic flow measurement (EFM) system for gas-gathering operations society of petroleum engineers (SPE). In: Paper presented at the Permian Basin Oil and Gas Recovery Conference, March 18–20, SPE-24001-MS (1992). https://doi-org.libproxy1.usc.edu/https://doi.org/10.2118/24001-MS

Jones, P.: Reservoir limit test on gas wells-. In: SPE-24, Gas Technology Symposium, Tyler, Texas (1962)

Katz, D.L., Lee, R.: In: Natural Gas Engineering production and storage, Mc Graw Hill Chemical Engineering Series (1990)

Rawlins, E.L., Schellhardt, M.A.: Backpressure data on natural gas wells and their application to production practices. U.S. Bureau of Mines, Monograph, vol 7. (1936)

Chapter 6
Naturally Fractured Reservoirs

Pressure transient testing in naturally fractured reservoirs presents unique challenges because of the complex flow behavior resulting from the presence of fractures in the reservoir. In naturally fractured reservoirs, the presence of fractures can significantly affect the flow behavior of fluids. The interaction between matrix flow and fracture flow needs to be understood and properly modeled during well testing analysis. Well testing in naturally fractured reservoirs often involves the use of dual porosity or dual permeability models. These models account for the separate flow mechanisms in the matrix and fractures, considering their distinct permeabilities and storage capabilities.

It is crucial to accurately identify and characterize the fractures in the reservoir before conducting well testing. Techniques such as wellbore imaging, and geophysical methods can help to determine fracture orientation, intensity, and connectivity. Well tests are essential for evaluating and characterizing the fractured reservoir. This includes identifying the presence of fractures, estimating their properties (e.g., fracture permeability, fracture porosity), and determining matrix properties. Interference testing can provide valuable information about the connectivity between fractures and the matrix. By monitoring pressure responses in different wells, the communication and heterogeneity of the fractured reservoir can be inferred. Using reservoir simulation models that incorporate fracture attributes can help in understanding the dynamic behavior of the reservoir and predicting future well performance. These models can be calibrated and validated with well testing data in naturally fractured reservoirs that require specialized analysis methods to account for the unique flow behavior and heterogeneity of the fractured system. Proper understanding of the fracture network, accurate data acquisition, and comprehensive interpretation techniques are crucial for effective reservoir management and production optimization.

There are different ways the presence of fractures can affect the pressure transient behavior of a naturally fractured reservoir. This is best explained by Nelson (1982) classification that differentiates among fractured reservoirs depending on the nature of flow. He distinguishes systems where flow enhancements are primarily attributed to

I. Ershaghi, *Solved Problems in Well Testing*,
https://doi.org/10.1007/978-3-031-47299-2_6

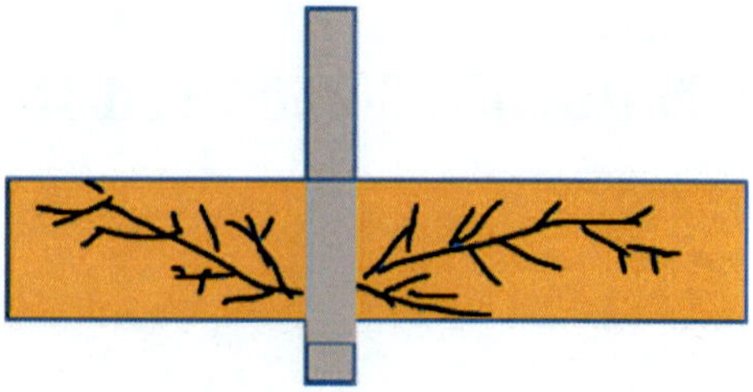

Fig. 6.1 Microfractures enhancing the interporosity flow

the presence of fractures in contrast to when the presence of fractures tends to enhance productivity in tight matrix conditions. In general, the actual pore volume of major fractures is not substantial. What matters the most is the presence of microfracture that facilitate the flow from matrix into the major fractures, Fig. 6.1.

Barenblatt et al. (1960) suggested the concept of double-porosity systems for describing flow in naturally fractured reservoirs. The system they discussed included a fractured reservoir consisting of two interacting, overlapping sub systems: a low-permeability matrix and a high-permeability, low-storativity fractured system. The fractures serve as a conduit for the fluid to flow to the well and the matrix supports the fractures system. Warren and Root (1963) focused on pseudo-steady-state matrix/fracture fluid transfer. In this system, a pressure gradient within the matrix does not exist and the matrix is treated as a time-dependent source term feeding the fracture depending on the fracture pressure. Kazemi (1979) considered transient flow in both matrix blocks and fractures. These approximation for using pseudo-steady-state and transient interporosity flow models have been discussed by several authors.

6.1 Matrix-Fracture Transfer Rate

In the model presented by Warren and Root (1963), two parameters, (ω) and (ʎ) are used to define the fracture-matrix system. These parameters define the contribution of the fractures to the overall system in terms of porosity and transmissibility.

Partition coefficient (ω) is defined as

$$\omega = \frac{\Phi m C m}{\Phi f C f + \Phi m C m}$$

where (m) refers to the rock matrix and (f) refers to the fracture system. The parameter (**ω**) may be estimated from the plot of pressure versus log time data, Fig. 6.2.

$$\omega = 10^{-\frac{(p_1 - p_2)}{m}} \tag{6.1}$$

where m is the slope of semi-log the straight lines in psi/cycle.

The inter-porosity flow coefficient parameter (λ) is defined as shown below:

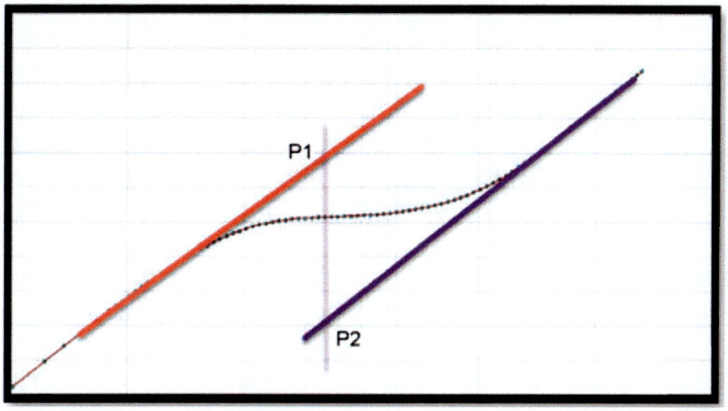

Fig. 6.2 Estimation of (ω) using the Warren and root model

$$\lambda = \frac{\alpha.km}{kf} rw^2 \tag{6.2}$$

Parameter (λ) is the second dual porosity parameter, which is also determined from the minimum point, (P'dt, min), on the pressure derivative curve using the procedure described by Deruyck et al. (1982). The parameter (λ) is quite small because matrix permeability is very low compared with fracture permeability. From pressure transient tests, the estimation of the inter-porosity flow coefficient as defined by lambda (λ) can be obtained in several ways. Various methods have been suggested. The interporosity flow coefficient (λ) may be evaluated using the following Equation:

$$\lambda = \frac{\omega}{t_{D,min}} \ln \frac{1}{\omega} \tag{6.3}$$

The work by Abdassah and Ershaghi (1986) resulted in an improved model for analysis of pressure transient tests of naturally fractured reservoirs. This was in terms observations of actual well tests that showed anomalous slope changes during the transition period and where dual-porosity models could not explain the behavior. They showed that the fracture-controlled early times and portions of the transition period resemble the behavior of a dual-porosity system and the transition zone in the latter part of, however, shows slope changes with the duration being a function of (λ1/λ2), ratio of interporosity flow coefficients, for the two matrix types) and (ω1/ω2) (ratio of fluid capacitance coefficients).

Alghamdi and Ershaghi (1996) maintained that dual fracture models are a more realistic alternative to dual porosity models for the representation of naturally fractured reservoirs. They showed that a component of the fracture system constituting microfracture responds later than the microfractures. They delineated microfracture response versus matrix response using the proposed conceptual models. They demonstrated that the microfractures response may at times be mistakenly attributed to matrix.

Rebolledo and Ershaghi (2015) showed the errors associated with the interporosity coefficient (λ), representing the communication between the matrix and fractures. As Warren and Root offered no solutions for estimation of (λ), they examined solutions that were proposed by several authors. That included a methodology suggested by the Gringarten (1984) method and that by Uldrich - Ershaghi (2015). They demonstrated that the application of some methods such the one suggested by Gringarten (1984) may result in errors of several orders of magnitudes in estimation of (λ). They emphasized various other techniques, including non-linear parameter estimations, Crawford et al. (1976) or the Uldrich-Ershaghi (1979) method, may be used to check the estimated values for λ.

The above discussions are all based on fractured reservoirs that can be represented by the dual porosity models discussed by Warren and Root (1963). The flow in such systems is considered pseudo steady estates or PSS. Alternative models have been introduced by Kazemi (1979) and Streslsova (1983) to represent fractured reservoirs that exhibit gradient flow and when reservoir systems are under unsteady state conditions, USS. The condition of unsteady state can develop when the matrix blocks are large, and a gradient flow from the matrix to fractures controls the flow system.

In the following examples we present the log–log pressure derivative plots versus time with a focus on the dual porosity PSS models, see Figs. 6.3, 6.4,6.5, 6.6, 6.7, 6.8, 6.9, 6.10, 6.11, 6.12, 6.13, 6.14, 6.15, 6.16, 6.17, 6.18, 6.19, 6.20 and 6.21.

Kazemi (1979) showed that analysis of a naturally fractured reservoir from pressure transient data relies on the degree and the type of heterogeneity of the system.

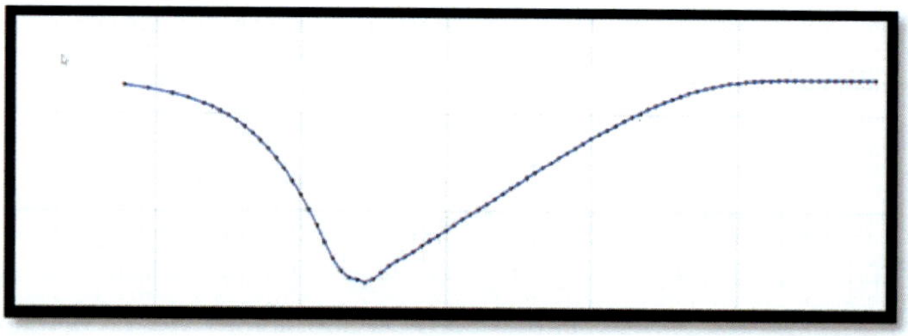

Fig. 6.3 Log–log pressure derivative function plot of NFR with no wellbore storage-skin effects

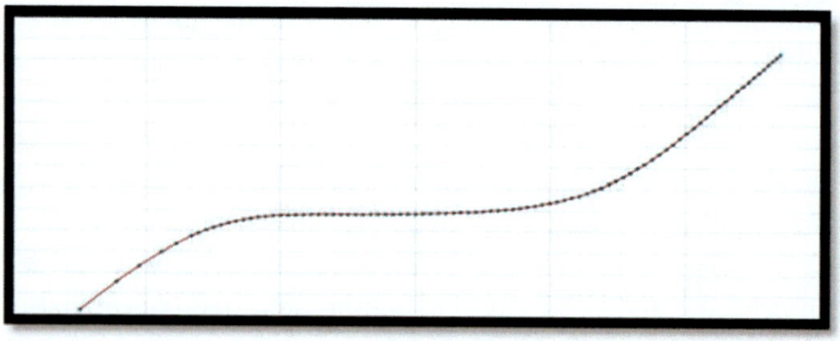

Fig. 6.4 Semi log plot of horner plot for a NFR without the effects of wellbore storage and skin

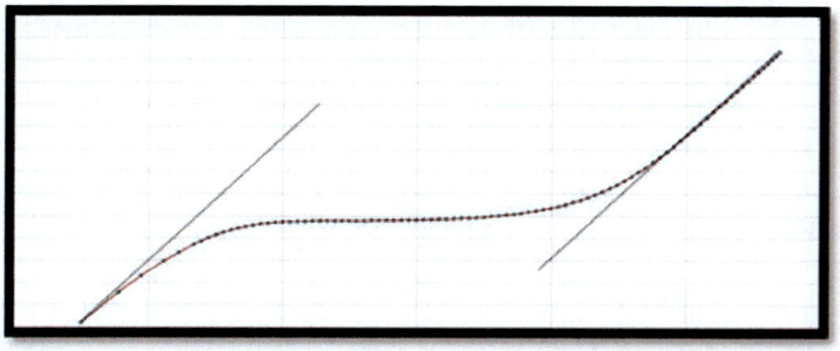

Fig. 6.5 Semi log plot of shut in pressure versus (t + dt)/dt for a NFR without the effects of wellbore storage and skin exhibiting two parallel straight lines

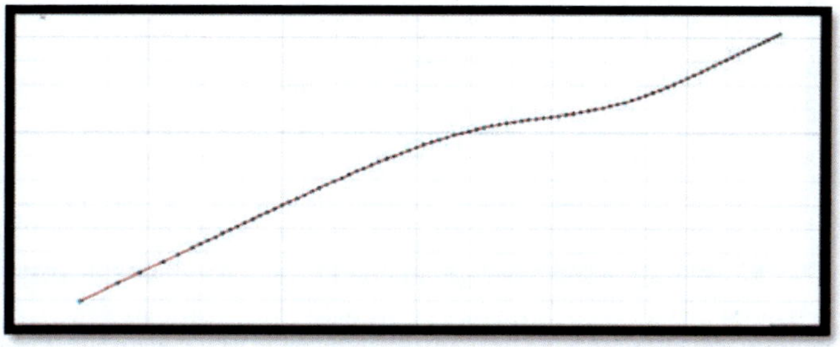

Fig. 6.6 Semi log plot of horner plot for a nfr without the effects of wellbore storage and skin with smaller contribution of matrix

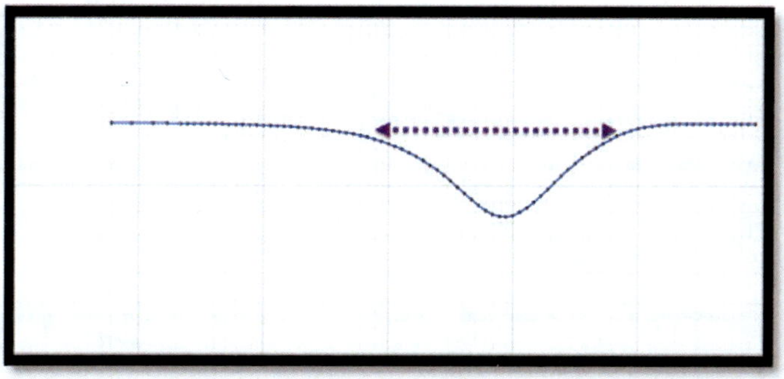

Fig. 6.7 Log-log pressure derivative function plot for a NFR without the effects of wellbore storage and skin with smaller contribution of matrix

Under favorable conditions, one can calculate in-situ characteristics of the matrix-fracture system, including the pore-volume ratio, over-all capacity of the formation, total storage capacity of the porous matrix, and some measure of matrix -fracture permeability ratio.

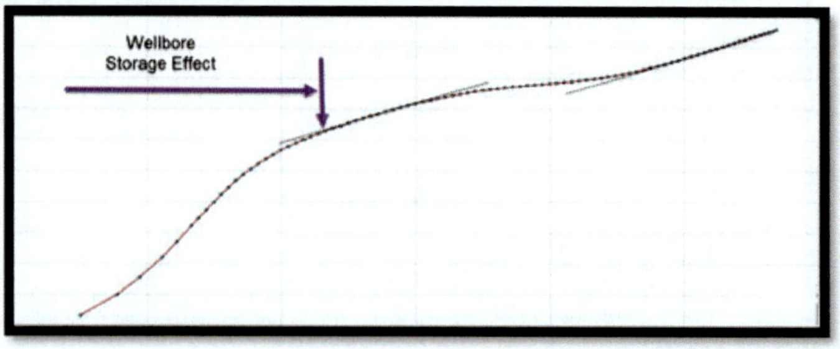

Fig. 6.8 Semi log plot of horner plot for a NFR with the effects of wellbore storage and skin

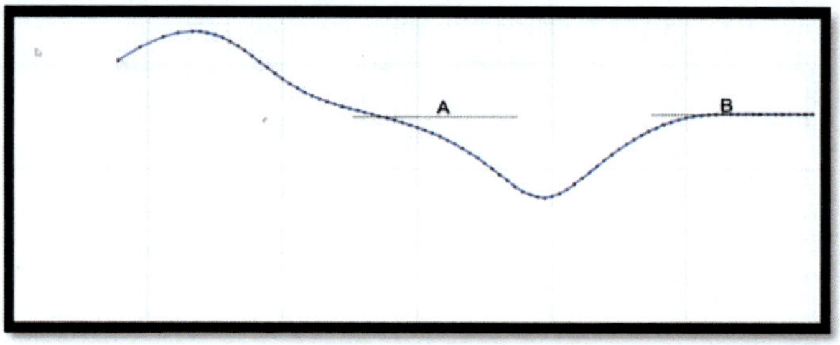

Fig. 6.9 Log-log pressure derivative function plot for a NFR with the effects of wellbore storage and skin with smaller contribution of matrix

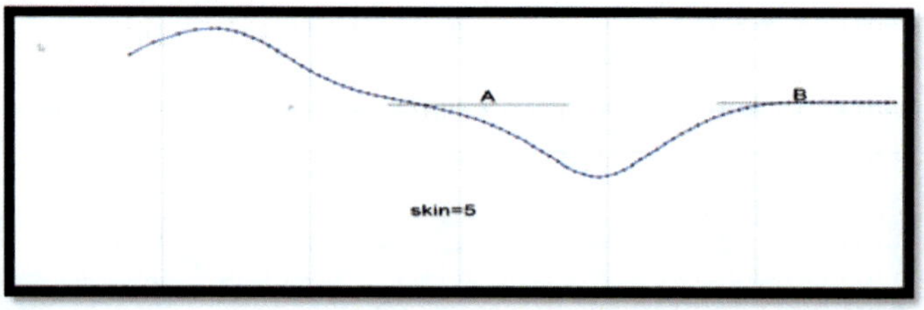

Fig. 6.10 Log-log pressure derivative function plot for a NFR with the effects of wellbore storage and skin = 5 with smaller contribution of matrix

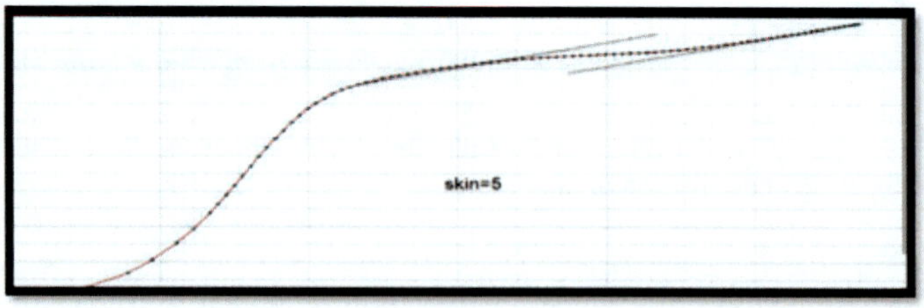

Fig. 6.11 Semi plot of pressure buildup for a NFR with the effects of wellbore storage and skin = 5

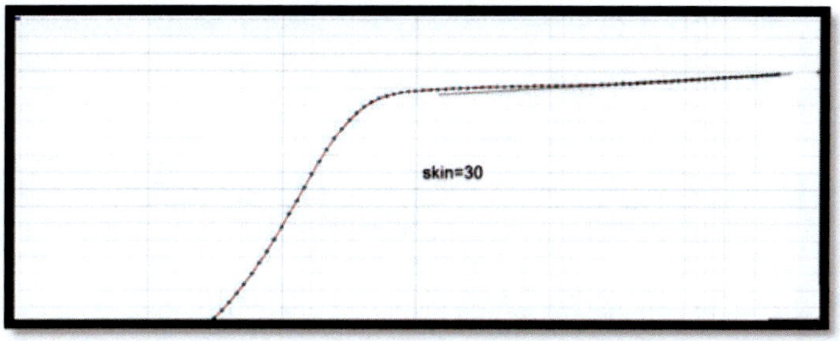

Fig. 6.12 Semi plot of pressure buildup for a NFR with the effects of wellbore storage and skin = 30

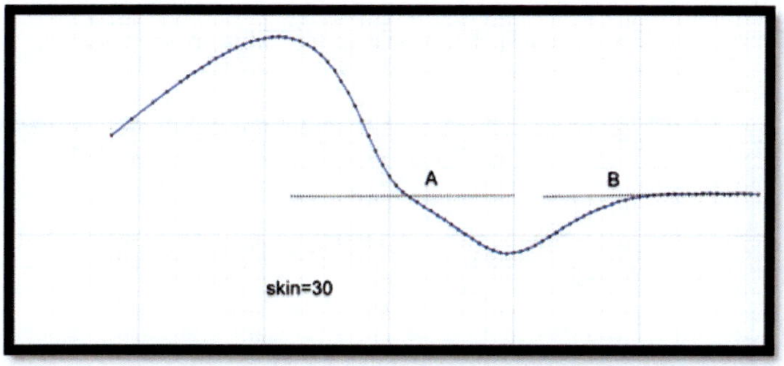

Fig. 6.13 Log-log of pressure derivative function plot for a NFR with the effects of wellbore storage and skin = 30

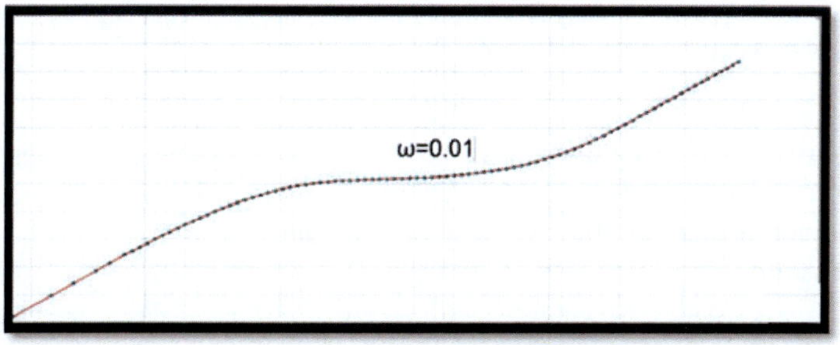

Fig. 6.14 Expanded semi-log plot for a dual porosity NFR with no wellbore storage and no skin, $\omega = 0.01$

6.2 Faults in NFR

While the subject of getting information about effective fracture properties and the type of interporosity flow are important, we need to realize that fractured reservoirs could be inundated with many geological faults. The expected fault response in NFR

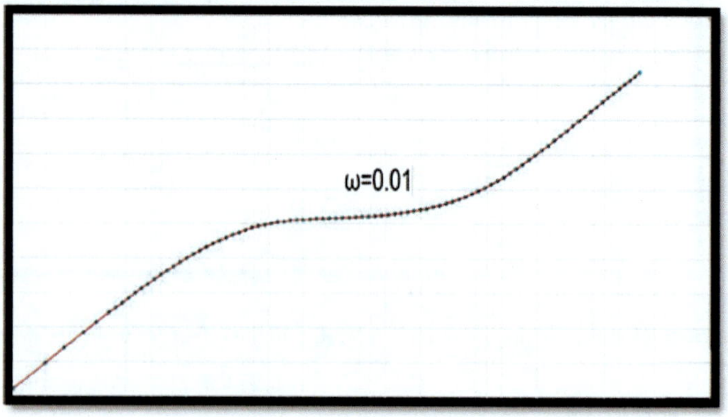

Fig. 6.15 Semi-log plot for a dual porosity NFR with no wellbore storage and no skin, $\omega = 0.01$

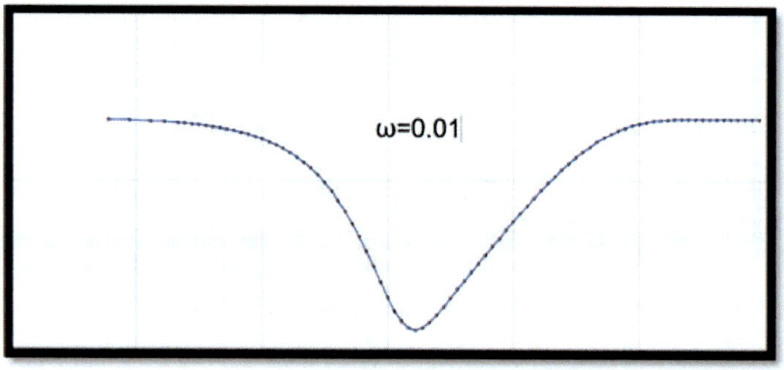

Fig. 6.16 Log-log plot of pressure derivative function plot for a dual porosity NFR with no wellbore storage and no skin, $\omega = 0.01$

was discussed by Khachatoorian et al. (1995) and as shown below predicted responses of the derivative plots depends on the fault distance to the well. Figure 6.22 shows the behavior of pressure transient response in dual porosity systems the presence of a sealing fault.

6.3 Questions and Answers

Q6.1: How do we estimate the percentage of pore volume occupied by fracturs in a naturally fractured reservoir?

A6.1: Estimating of (ω) can provide a basis for such estimation.

Q6.2: How can we estimate the interporosity coefficient?

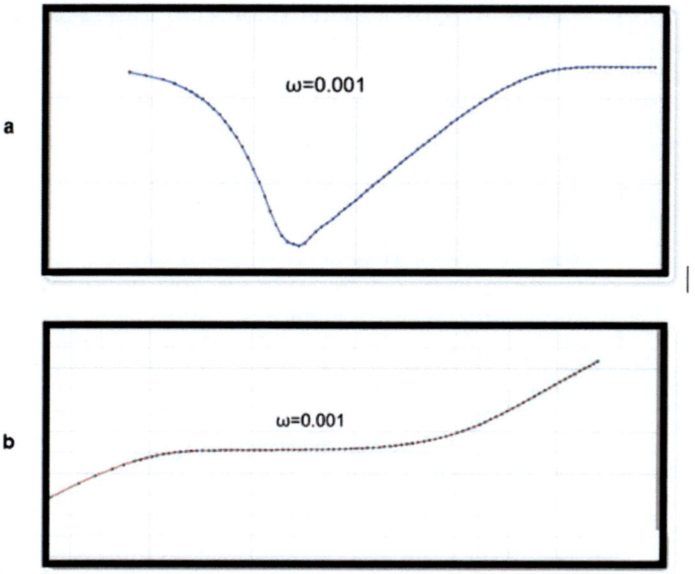

Fig. 6.17 **a** Log-log plot pressure derivative function plot and **b** Semi-log plot and for a dual porosity NFR with no wellbore storage and no skin, $\omega = 0.001$

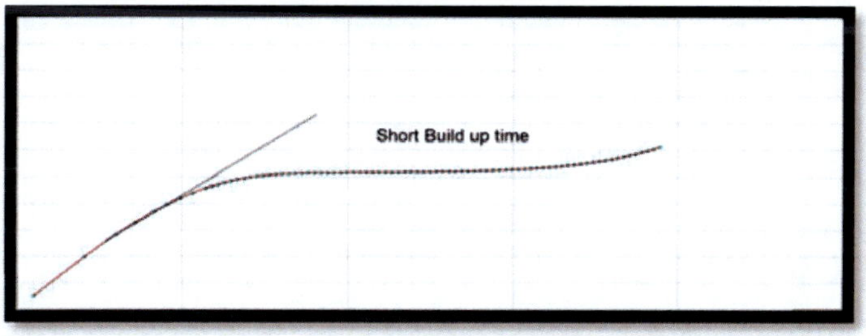

Fig. 6.18 Semi-log plot of for a dual porosity NFR with no wellbore storage and no skin, with a very small λ

A6.2: Issues in estimation of (λ) is discussed by Robolledo and Ershaghi (2015).

Q6.3: How can we identify the relative block sizes?

A6.3: Small matrix block sizes generate a pseudo steady state response on a buildup analysis with a clear point of inflection seen on the transition segment.

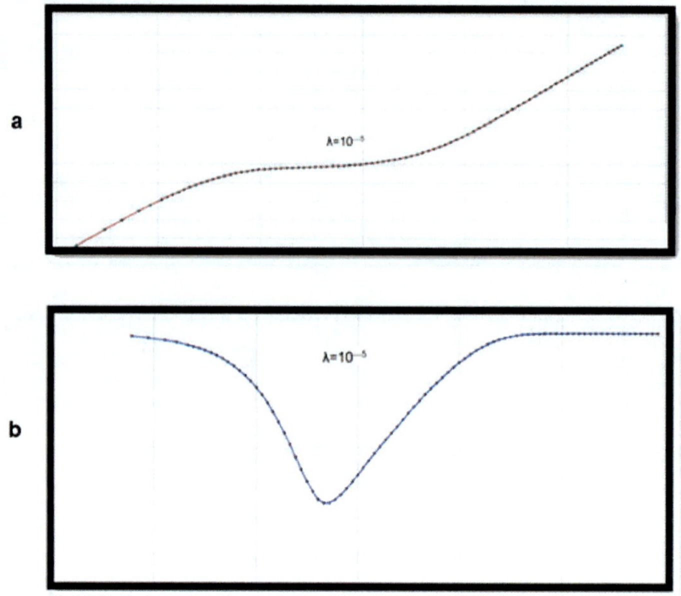

Fig. 6.19 **a** Semi-log plot of pressure buildup and **b** Log–log plot of the pressure derivative function plot for a dual porosity NFR with no wellbore storage and no skin, with $\lambda = 10^{-5}$

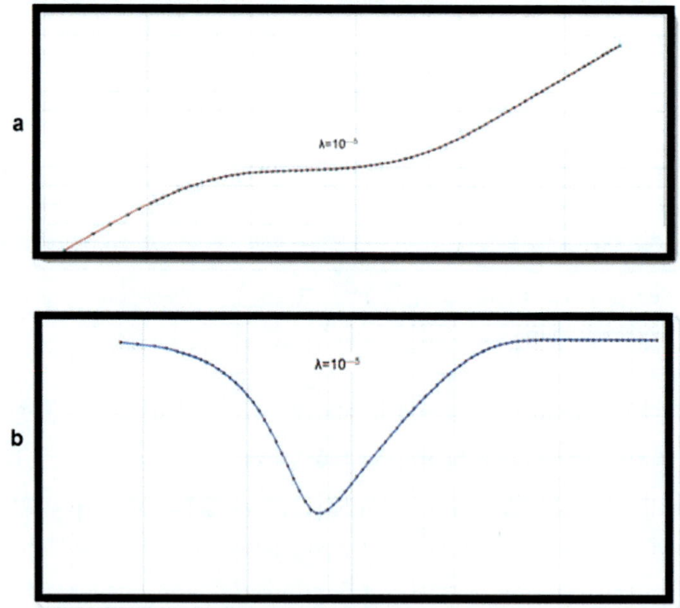

Fig. 6.20 **a** Semi-log plot of pressure buildup and **b** Log-log plot of the pressure derivative function plot for a NFR with no wellbore storage and no skin, with $\lambda = 10^{-6}$

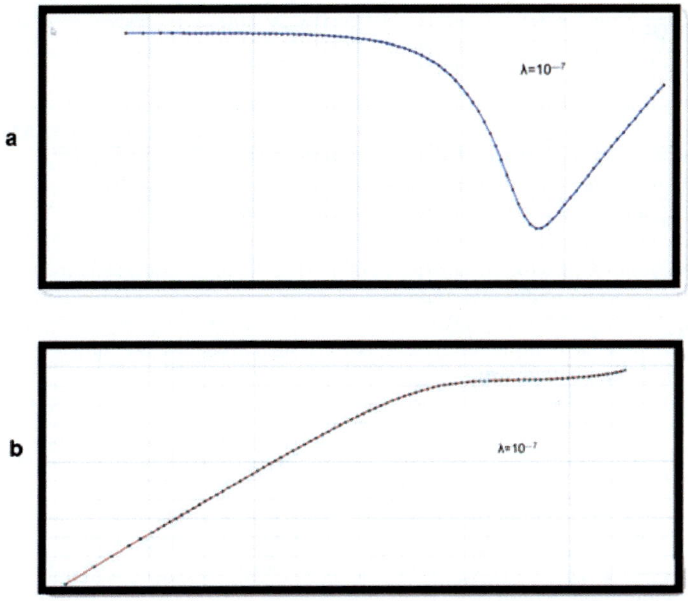

Fig. 6.21 Semi-log plot of pressure buildup and log a log plot of the pressure derivative function plot for a NFR with no wellbore storage and no skin, with $\lambda = 10^{-7}$

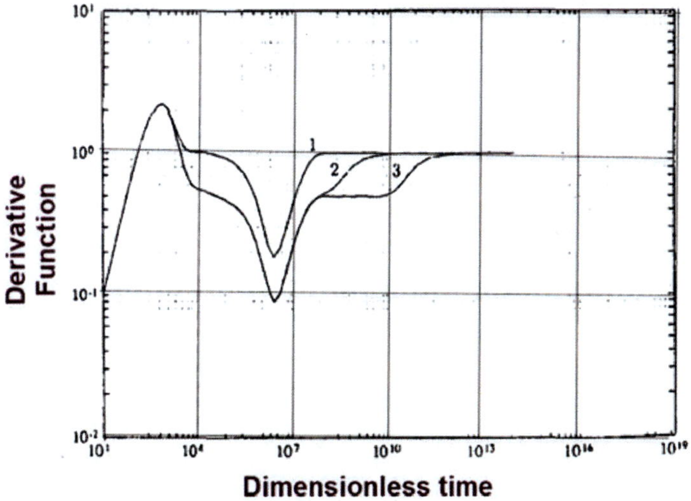

Fig. 6.22 Pressure derivative function plot for a sealing fault in naturally fractured reservoirs. (Khachatoorian et al. 1995). (1, 2 and 3 refer to incresaing the distance of the fault to the well)

References

Doddy, A., Iraj, E.: Triple-porosity systems for representing naturally fractured reservoirs. SPE Form Eval. **1**, 113–127 (1986). https://doi-org.libproxy2.usc.edu/10.2118/13409-PA

Abdullah, A.-G., Iraj, E.: Pressure transient analysis of dually fractured reservoirs. SPE J. **1**, 93–100 (1996). https://doi-org.libproxy1.usc.edu/https://doi.org/10.2118/26959-PA.

Barenblatt, G.I., Zheltov, I.P., Kochina, I.N.: PMM (soviet Applied Mathematics and Mechanics) **24**, 852 (1960)

Crawford, G.E., Hagedorn, A.R., Pierce, A.E.: Analysis of pressure buildup tests in a naturally fractured reservoir, SPE-AIME (1976)

Deruyck, B.G., Bourdet, D.P., DaPrat, G., Ramey, H.J.: Interpretation of interference tests in reservoirs with double porosity behavior-theory and field examples SPE-11026.MS (1982). https://doi.org/10.2118/11026.MS

Gringarten, A.C.: Interpretation of tests in fissured and multilayered reservoirs with double-porosity behavior: theory and practice, SPE (1984)

Kazemi, H.: Pressure transient analysis of naturally fractured reservoirs with uniform fracture distribution. Soc. Petrol. Eng. J. **9**(04), 451–462 (1969). https://doi.org/10.2118/2156-A

Khachatoorian, R., Ershaghi, I., Shikari, Y.: Complexities in the analysis of pressure-transient response in faulted naturally fractured reservoirs. SPE Form. Eval. (1995)

Nelson, R.A: An approach to evaluating fractured reservoirs. J. Pet. Technol. Soc. Petrol. Eng. (SPE) **34**(09), 2167–2170 (1982)

Rebolledo, L.C., Ershaghi, I.: Errors in estimating interporosity flow coefficient in naturally fractured reservoirs. In: Paper presented at the SPE Annual Technical Conference and Exhibition, Houston, Texas, USA, September (2015). https://doi-org.libproxy2.usc.edu/https://doi.org/10.2118/178737-STU

Streltsova, T.D.: Well pressure behavior of a naturally fractured reservoir. Soc. Petrol. Eng. J. SPE J **23**(05), 769–780 (1983). Published: 01 Oct 1983

Uldrich, D.O., Iraj, E.: A method for estimating the interporosity flow parameter in naturally fractured reservoir. SPE-AIME (1979)

Warren, J.E., Root, P.J.: The behavior of naturally fractured reservoirs. SPE J **3**(03), 245–255 (1963). Published: 01 Sept 1963

Chapter 7
Hydraulically Fractured Wells

When a formation is tight with low permeability, breaking the formation by creating a hydraulic fracture can help to increase the production rate from that formation. Initial developments related to the tight sands. In these cases, the goal has been to develop primarily one major fracture in the target zone. With the advent in the development of source rocks (unconventional resources), a lateral horizontal well is drilled, and then multiple stages of fracturing are engineered.

Diagnostic Fracture Injection Tests (DFIT) are commonly used in fractured reservoirs to assess fracture closure pressure, minimum in-situ stress, and fracture properties. These tests involve injecting small volumes of fluid at different rates to evaluate fracture behavior and to estimate key parameters.

There are several techniques to analyze transient pressure data from hydraulically fractured wells. They include modified pseudo-radial flow (semi-log analysis), formation linear flow (square-foot time analysis), bilinear flow (fourth-root time analysis) using type curve analysis, and reservoir simulation history matching analysis. By these techniques we can evaluate formation matrix permeability, fracture half-length and fracture conductivity. Each technique has its advantages and disadvantages in estimating the above-mentioned parameters.

Well tests can help to estimate the effectiveness of fractures and their conductivities. For fractured formations, the nature of the flow regime changes to a linear system. Detection of the linear flow system and the extent of the fractured wings can be obtained from well tests data. When a well is hydraulically fractured, effective flow properties of fractures may be estimated. This includes fracture conductivity, fracture width and fracture length, Fig. 7.1.

As mentioned before, with the existence of a hydraulic fracture, the nature of the flow changes and becomes of a linear flow type. A Horner plot of pressure buildup tests exhibits a negative skin followed by radial flow, Fig. 7.2.

But the derivative function log–log plot can readily exhibit the existence of linear flow as a straight line with a slope of 0.5, Fig. 7.3.

I. Ershaghi, *Solved Problems in Well Testing*,
https://doi.org/10.1007/978-3-031-47299-2_7

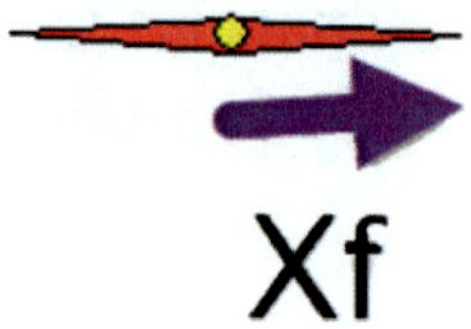

Fig. 7.1 The definition of the wing of a hydraulic fracture

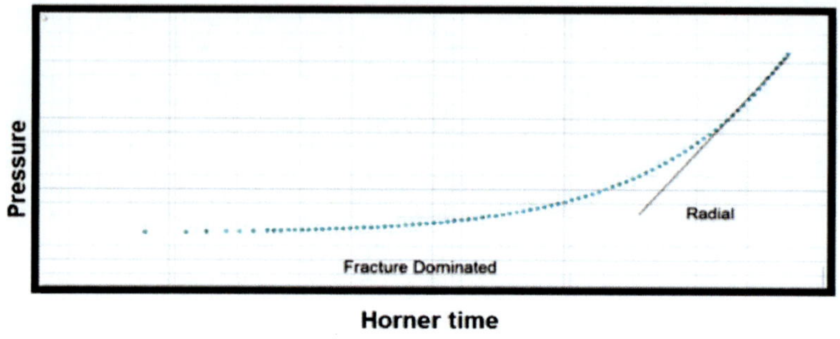

Fig. 7.2 A typical semi log plot of build up pressure versus log of horner time

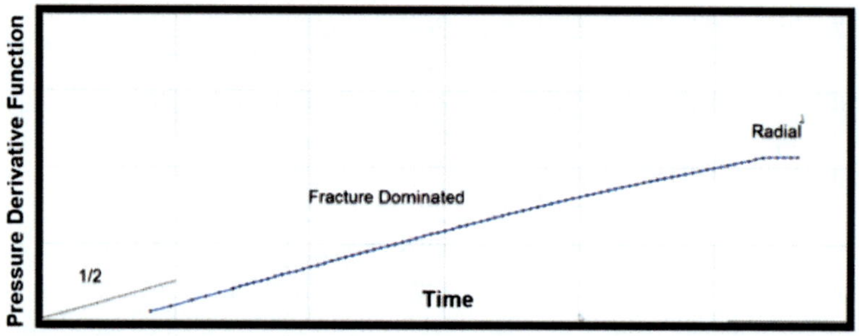

Fig. 7.3 A typical log-log plot of the pressure derivative function plot exhibiting linear flow caused by the presence of a fracture

Wells subjected to hydraulic fracturing can show a substantial increase in their productivity. The important thing is making sure the fracture efficiency has been optimal. The reason for a slope of 0.5 on a log-log plot of pressure derivative function plot vs. time is the basic equation relating pressure to time. As shown below, it can be shown, the pressure changes are a square root function of time:

$$Pd = \alpha \cdot \sqrt{td} \tag{7.1}$$

$$\log(\mathrm{Pd}) = \log(\alpha) + 0.5^{*} \log(\mathrm{td}) \tag{7.2}$$

where (α) is a constant. But the slope of 0.5 is also based on the assumption that the fracture conductivity is very high. Fractures that are of low conductivity can

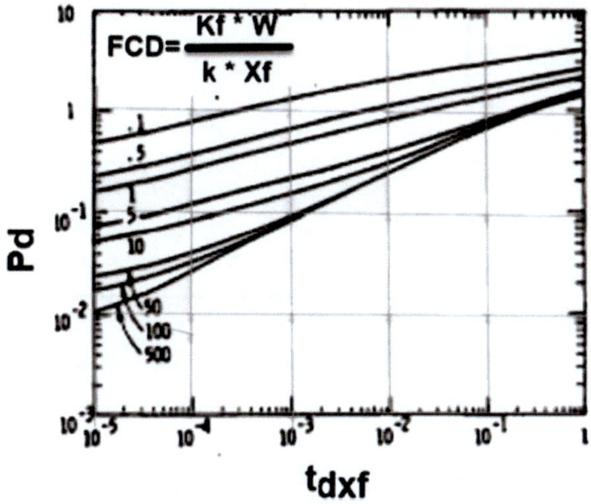

Fig. 7.4 The type curve presented by Agarwal et al. 1979

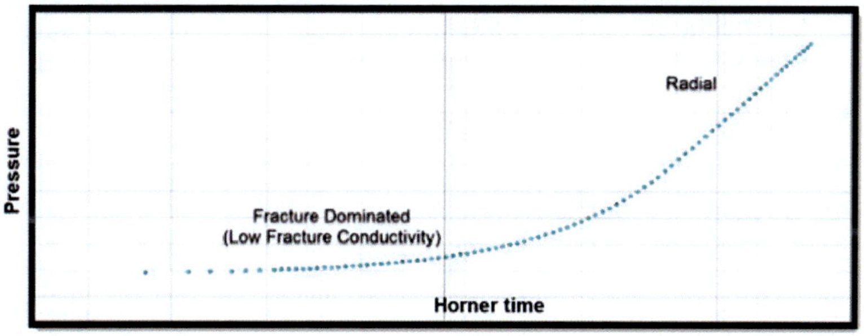

Fig. 7.5 A typical semi log plot of build up pressure versus log of horner time for low conductivity fracture

show a reduced slope on a log plot of dimensionless pressure vs. dimension time. As shown by Agarwal et al. (1979), Fig. 7.4, on a long-time, the pressure vs. time, the manifestation of low fracture conductivity is a slope developing less than 0.5.

As shown in Figs. 7.5 and 7.6, both the plot of pressure changes with log of Horner time and the derivative function lots can be diagnostics.

The lower the value of fracture conductivity, the less efficiency can be attributed to the fracture job. For cases where fracture conductivity is low and the observed slope on the log–log plot is less than 0.5, examination of the causes is in order to improve future fracturing jobs. As shown in Fig. 7.7, besides factors related to the permeability drops caused by the placement of proppants, the fracture job itself and the brittleness of the rock can impact the fracture development and fracture conductivities.

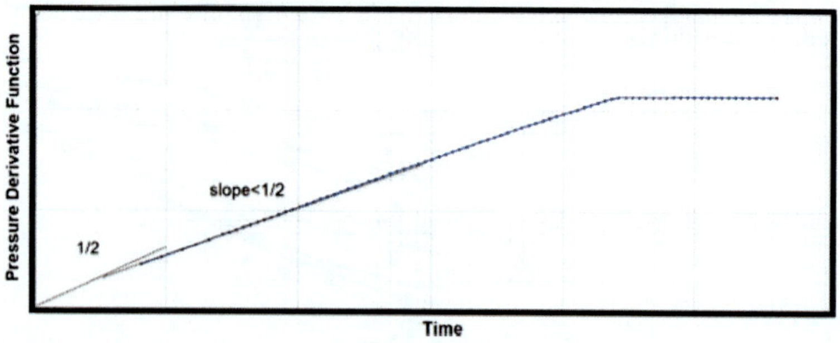

Fig. 7.6 A typical log -log plot of the pressure derivative function plot exhibiting low fracture conductivity

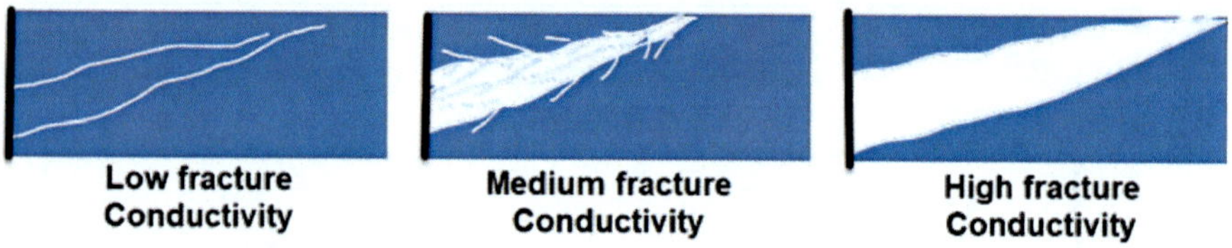

Fig. 7.7 Some fracture development patterns

Fig. 7.8 The graphical representation of fracture skin

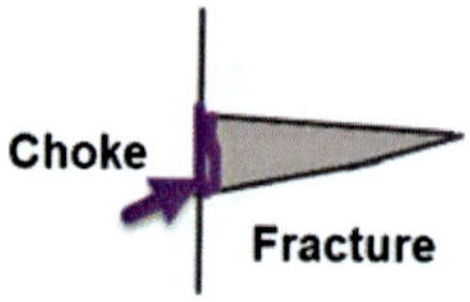

7.1 Estimation of Fracture Skin

Fracture permeability may not be uniform along the fracture length and if particularly around the producing well the fracture is choked by proppants, the evidence of low fracture conductivity can be detected, Fig. 7.8.

7.2 Questions and Answers

Q7.1: How can we estimate the fracture conductivity?

A7.1: We can use a type curve such as the one published by Agarwal et al. (1979) and by matching the data with the proper trend on the type curve and estimating the fracture conductivity.

Q7.2: How can we estimate the average fracture width?

A7.2: By matching the data with a trend we can obtain fracture conductivity and from that we can obtain the fracture width.

Q7.3: How can we recognize fracture-controlled transients?

A7.3: Fracture controlled transients exhibit non radial flow and mostly linear flow conditions.

Q7.4: How can we compute the fracture length?

A7.4: We can use a type-curve such as the one published by Agarwal et al. (1979) and by matching the data with the proper trend on the type of curve and estimate fracture length from the time match.

Q7.5: How can we compute the fracture length?

A7.5: We can use a type of curve such as the one published by Agarwal et al. (1979) and by matching the data with the proper trend on the type of curve and estimate fracture length from the time match.

Reference

Agarwal, R.G., Carter, R.D., Pollock, C.B.: Evaluation and performance prediction of low-permeability gas wells stimulated by massive hydraulic fracturing. J. Petrol. Technol. **31**(03), 362–372 (1979). https://doi.org/10.2118/6838-PA

Chapter 8
Horizontal Wells

This will help to gather data about reservoir properties such as reservoir pressure, permeability, and skin damage and will help in optimizing production and reservoir management.

During the test, the flow rate is typically changed by adjusting the surface chokes. The response of the pressure over time is analyzed to determine the reservoir properties using various well testing analysis techniques. Horizontal wells offer unique challenges during pressure transient testing because of their complex flow behavior. Depending on the well completion design and reservoir characteristics, the pressure response can exhibit radial, pseudo-radial, or linear flow regimes. Analysis methods such as type curve matching and numerical simulations are employed to interpret the test data and estimate reservoir parameters.

The best way to understand the behavior of horizontal wells is to examine the pressure derivative function vs, time on a log–log plot. As shown in Fig. 8.1 for a long duration test, three periods may be recognized. The first part represents the vertical radial flow, second represents a linear flow period and finally, a second radial flow represents the flow from the reservoir drainage area into the wellbore.

For horizontal wells, there are solutions presented in the literature that allow estimation of reservoir permeability. Formulations are available to estimate permeability in three different directions. These formulations assume the effective length of the horizontal lateral is known. If the well is not an actual horizontal well and is drilled with ups and downs and a nonlinear shape, then the effective length of the well is also an unknown and needs to be estimated.

A study done by Qi et al. (2017) focused on the challenge of additionally obtaining the effective length of the lateral in horizontal wells. In this study it was pointed out that horizonal wells may not navigate within a given target as planned and the navigation path may result in having a less effective length. Adding this unknown to the list of unknown permeability requires using an iterative method for solving the resultant nonlinear equations.

I. Ershaghi, *Solved Problems in Well Testing*,
https://doi.org/10.1007/978-3-031-47299-2_8

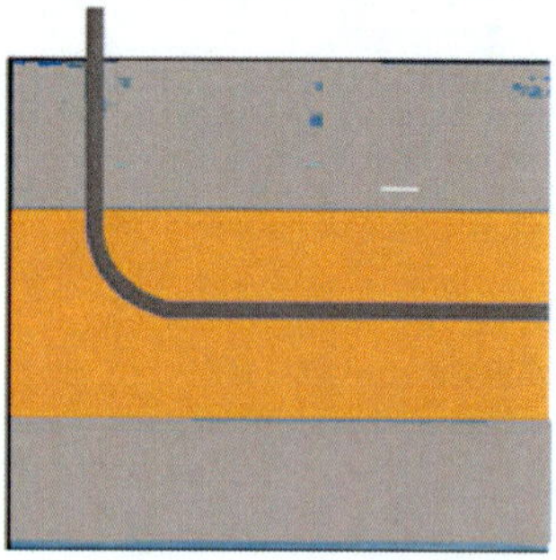

Fig. 8.1 Schematic of a horizontal well

8.1 Questions and Answers

Q8.1: How do we estimate the permeability in three dimensions from a horizontal well test?

A8.1: We can estimate the permeabilities in x, y, and z directions by solving three equations early radial flow: The semi-log plot of the initial radial flow period

$$p_{ws} = p_i - \frac{162.6q.\mu.B}{\sqrt{kx.kz.Lw}} \log \frac{t+dt}{dt} \tag{8.1}$$

where a plot of Pws versus log $(\frac{t+dt}{dt})$ can exhibit a straight line and the slope include two unknowns kx and kz can be obtained. See Good and Thambynayagam (1987) and Odeh and Babu (1990), and Kuchuk et al. (1998).

After the radial flow, a linear flow period will develop where a Cartesian plot of Pws versus $(\sqrt{t+dt} - \sqrt{dt})$ results in another equation that includes kx in the slope of the plot.

$$Pws = Pi - \frac{8.12q.B}{hz.Lw}\sqrt{\frac{1}{k_x.\varphi.c_t}}\left\{(\sqrt{t+dt} - \sqrt{dt})\right\} \tag{8.2}$$

Finally, the late radial flow period provides another equation between Pws and log (t) that includes kx and ky

$$Pws = Pi - \frac{162.6q.\mu.B}{h\sqrt{k_x.k_y}} \log \frac{t+dt}{dt} \tag{8.3}$$

We can also use equations such as the intersection points of the radial flow and linear flow or end of linear for permeability estimation purposes.

Q8.2: How can we estimate the effective length of a horizontal well?

A8.2: If the well is not truly a horizontal well and the actual length represents the model of a snake-shaped navigation path, the length will be unknown and that estimation together with Kx, Ky and K_Z constitute a 4th unknown. See Qi et al.

Q8.3: How can we estimate skin around a horizontal well?

A8.3: Skin around a horizontal well exhibits itself in the early radial flow period and the estimation can be using equation such as the one shown below:

References

Good, P.A., Thambynayagam, R.K.M.: Pressure drawdown and buildup analysis of horizontal wells in anisotropic media. SPE 14250 (1987)

Kuchuk, F et al.: Performance evaluation of horizontal wells (Schlumberger and GeoQuest). SPE 49539 (1998)

Odeh, A.S., Babu, D.K.: Transient flow behavior of horizontal wells: pressure drawdown and buildup analysis. SPE-18802-PA (1990)

Qi, Q., Bethel, K., Ershaghi, I.: On estimation of horizontal well effective lateral length and reservoir properties from well test data. In: Paper Presented at the SPE Western Regional Meeting, Bakersfield, California, April 2017. Paper Number: SPE-185645-MS (2017). https://doi.org/10.2118/185645-MS

Chapter 9
Pulse and Interference Tests

Pressure pulse and interference testing are methods to evaluate the connectivity and communication among wells in a subsurface reservoir. These tests involve injecting or producing fluid from one well and monitoring the pressure response in surrounding wells. In pressure pulse tests, a pressure pulse is generated by briefly shutting in or opening a well. The resulting pressure wave propagates through the reservoir, and the pressure response is measured in other wells. By analyzing the arrival time and amplitude of the pressure pulse in different wells, we can determine the transmissibility and storativity between pulsing wells and detecting reservoir boundaries. Interference testing, on the other hand, involves monitoring the pressure response in multiple wells while one well is being produced or subjected to injection. The pressure changes observed in the adjacent wells help in identifying the communication pathways and quantify the reservoir properties, such as transmissivity and storativity.

These tests are useful in assessing reservoir heterogeneity, hydraulic connectivity, and well-to-well communication. They provide valuable information for reservoir characterization, well placement, and optimizing production or injection strategies.

The interpretation of pressure pulse and interference test data requires modeling and analysis techniques to account for fluid flow dynamics in the reservoir. These tests, combined with other reservoir monitoring and data acquisition methods, assist in making decisions regarding reservoir management.

Since Elkins and Skov (1946) introduced the concept of interference test to the petroleum industry, many developments in interpretation techniques have been made and various applications have been developed to characterize reservoirs away from the penetration points.

Interference and pulse testing are for the estimation of interwell reservoir properties such as porosity and permeability by first estimating (T), transmissibility and (s), storativity. The flow rate at the active well is changed periodically between flowing and shut-in while the pressure is monitored at an observation well some distance away. In the interference test the goal is estimating pressure responses at the observation wells. In pulse test analysis, we draw a tangent to the pressure–time curve

I. Ershaghi, *Solved Problems in Well Testing*,
https://doi.org/10.1007/978-3-031-47299-2_9

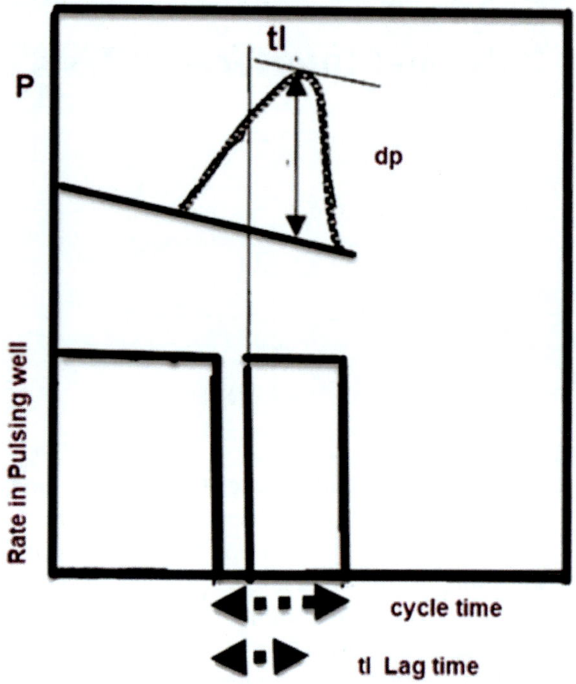

Fig. 9.1 Estimation of time lag (Tl) and pressure pulse, dp

between two peaks or valleys and a parallel tangent at the valleys (or peaks). Time lags and the pressure response amplitude are then estimated. Correlation curves developed by Kamal and Brigham (1975) for unequal pulse and shut-in periods are used to estimate the reservoir transmissibility (T) and storativity (s). In these correlation curves, only the time lag (t_{lag}), Fig. 9.1 is used. Discrepancies in the original Kamal-Brigham curves have led to new correlation curves.

The correlation curves that relate the relative time lag and dimensionless response amplitude to the dimensionless cycle period are obtained by solving the three equations that specify the equality of the slopes of tangents at the three-time lags with the slope of the line connecting the two outer points. Expressions for the pressures and pressure derivatives are used in these equation's solution with the application of the principle of superposition. Figure 9.1 shows the meaning of time lag and the typical correlation plots shown in Figs. 9.2 and 9.3 can be used to estimate dimension pressure and time values for the calculation of T and S as shown in Fig. 9.4. Parameter F is the ratio of pulse time to the cycle time.

$$R' = \frac{LagTime}{CycleTime}$$

Interference tests help provide information on the dynamic behavior of the reservoir including its connectivity and the existing permeability anisotropy, Pan et al.

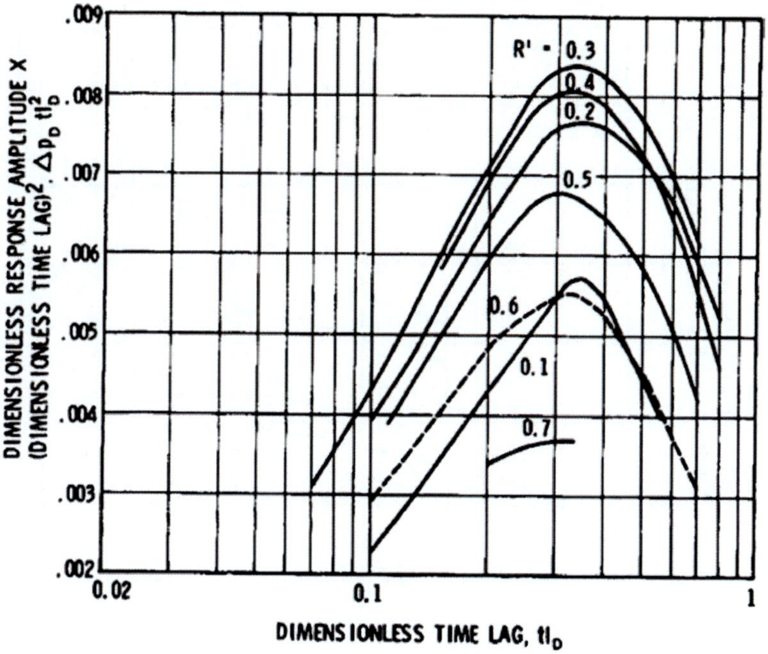

Fig. 9.2 Pulse test correlation plot for the first even pulse or estimation of T. (Courtesy Kamal and Brigham 1975)

(2019). It is important to realize that the information obtained, and the pressure match depends on appropriate test design. In complex fractured reservoirs matrix and fractured segments can be characterized by interference test results. Getting multiple interwall interference tests can help to obtain important information that will significantly help the reservoir characterization efforts particularly in complex heterogeneous, compartmentalized, naturally fractured and anisotropic reservoirs.

9.1 Use of Interference Tests

As indicated before, interference tests can help in detecting communication or lack of communication among wells. Communication may be via faults, fracture, and high permeability channels. For example, communication between horizontal wells is often seen because of induced fractures in unconventional resources. For unconventional wells, communication can be a secondary effect because of fracturing induced on individual wells. Such communication has been reported on some unconventional plays such as Bakken and Anadarko Basin in the U.S. Establishment of the presence of interference among horizontal wells is important as it can affect the economic life of wells.

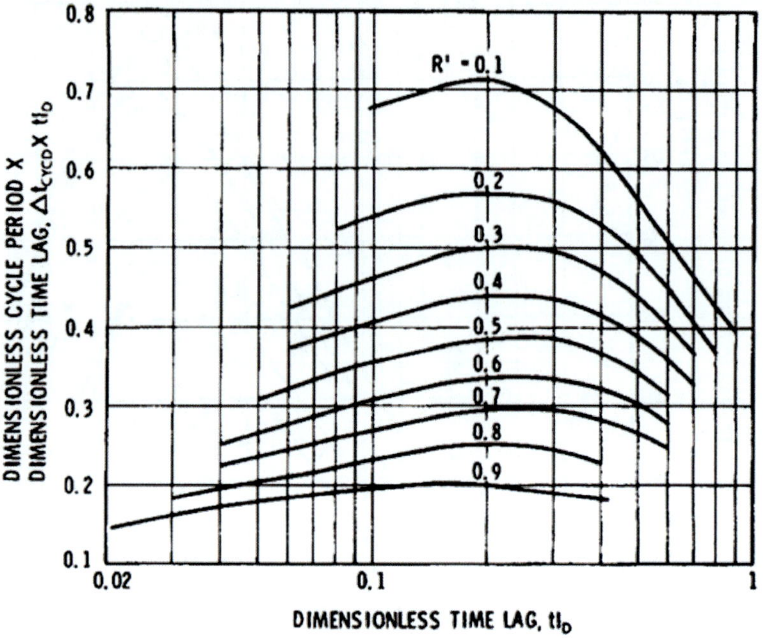

Fig. 9.3 Pulse test correlation plot g for the first even pulse to estimate S. (Courtesy Kamal and Brigham 1975)

Fig. 9.4 Estimation of T and S from pulse tests

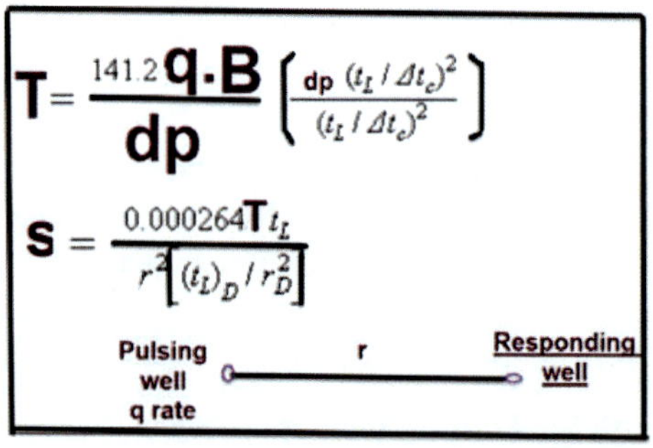

References

Elkins, L.F., Skov, A.: Determination of fracture orientation from pressure interference-transactions of the AIME **219**(01), 301–304 (1946).https://doi.org/10.2118/1516-G

Kamal, M., Brigham, W.E.: Design and analysis of pulse tests with unequal pulse and shut-in periods. J. Petrol. Technol. **28**(02), 205–212 (1975). https://doi.org/10.2118/4889-PA-1975

Pan, Y., Kamal, M., Narr, W.: Fieldwide determination of directional permeabilities using transient well testing. SPE Res. Eval. Eng. **22**, 734–744 (2019). https://doi-org.libproxy2.usc.edu/https://doi.org/10.2118/181437-PA

Chapter 10
Wells in Unconventional Reservoirs

Unconventional reservoirs producing from shale gas and shale oil make important contributions to hydrocarbon production capacity particularly in North America. Recovery from these reservoirs is via horizontal wells with multiple transverse fractures. A challenge in these types of reservoirs is estimating the effective productive volume created during stimulation and quantifying the expected ultimate recovery.

The development of low permeability gas reservoirs requires effective hydraulic fractures. For such reservoirs, we first observe the initial rapid decline rates. An extremely rapid decline rate is a sign of an unsuccessful stimulation treatment, Fig. 10.1. Such poor performance can be caused by extremely low formation permeability, poor fracturing operations, insufficient fluid in place and issues causing fracture permeability drops by crushing or embedment of propping agent.

In recent decades, the development of unconventional resources has opened up opportunities to drill and complete horizontal wells to access these tight rocks. To create the maximum contact, these horizontal wells are completed with multistage fracturing, Fig. 10.2. The number of fracture stages in horizontal wells tapping these unconventional resources has increased and can exceed 50 or more stages at spacings that are on other order of 200 ft.

In the development of such low permeability reservoirs, effective hydraulic fractures are needed. Because of low permeability of formation rock, rapid decline rates in such reservoirs can be seen sometimes even when a well has been successfully stimulated with hydraulic fractures.

Well testing in unconventional reservoirs plays a crucial role in assessing reservoir productivity, optimizing completion strategies, and providing key data for reservoir characterization and production forecasting. Shale gas or tight oil formations, presents unique challenges due to their low permeability and complex flow behavior. Unconventional reservoirs require hydraulic fracturing to enhance reservoir productivity. Well testing is typically performed after hydraulic fracturing to evaluate the effectiveness of the stimulation and fracture design.

I. Ershaghi, *Solved Problems in Well Testing*,
https://doi.org/10.1007/978-3-031-47299-2_10

Fig. 10.1 Typical production response from an unconventional play

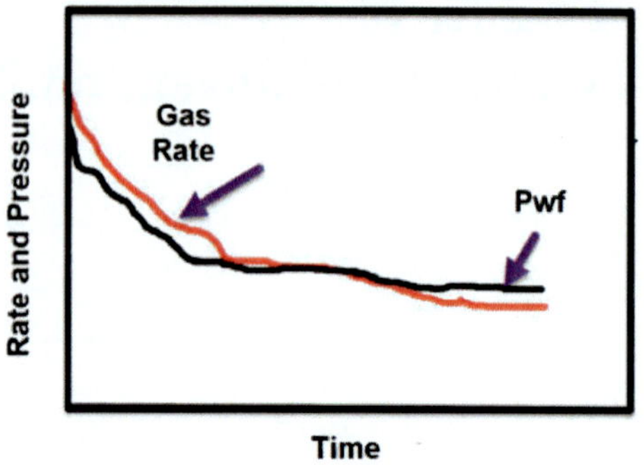

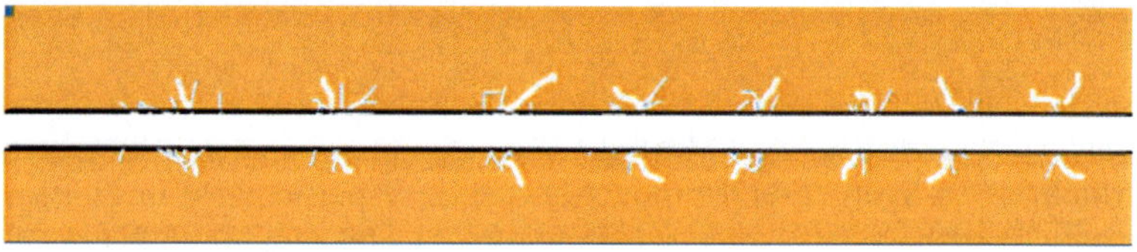

Fig. 10.2 Fracture distribution along the horizontal length

These reservoirs are developed using multi-stage fracturing techniques, where multiple fractures are created along the horizontal wellbore. The well testing may include isolating certain stages or clusters to evaluate their individual contributions to production.

These reservoirs exhibit transient flow behavior, where fluid flow rates change rapidly due to the low permeability and fluid storage within the matrix. In transient well testing methods, rate transient analysis, are often employed to estimate reservoir properties and production forecasting. Besides transient testing, long-term flow tests may be conducted to obtain data for reservoir characterization and for a better understanding of production behavior over an extended period. This can include continuous or intermittent flow rate monitoring.

DTS technology can be used to measure temperature profiles along the wellbore during well testing. One may identify preferential flow paths, understand reservoir heterogeneity, and monitor fluid distribution. PVT analysis is crucial for understanding the fluid properties of unconventional reservoirs, including gas-oil ratio, fluid composition, and phase behavior. PVT data obtained from well testing is critical for reservoir simulation and production forecasting. Comprehensive data integration and analysis techniques are employed to interpret well-test data from unconventional reservoirs. This includes combining wellbore pressure measurements, production data, and other measurements to estimate reservoir properties, optimize completion and stimulation strategies, and make a forecast of future production performance.

Unconventional reservoirs often require large volumes of water for hydraulic fracturing. Well testing should comply with environmental regulations to ensure proper water management and protection of the surrounding environment.

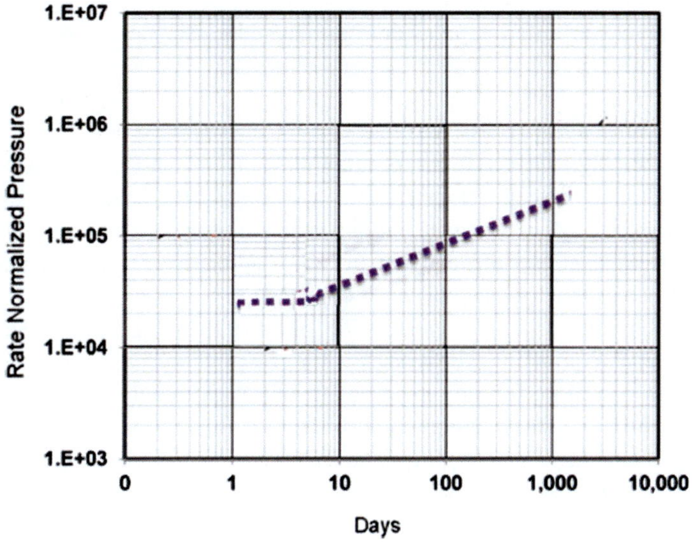

Fig. 10.3 Rate normalized plot for producing well from an unconventional reservoir

There are benefits of obtaining buildup responses whenever the well is shut-in for operational reasons. This can allow obtaining important insights about heterogeneity and compartmentalization in such reservoirs. If the initial reservoir pressure is unknown or pressure data accuracy is questionable, ambiguous results can be generated using deconvolution method.

But regular measurements of daily production rates and wellhead flowing pressures can provide some very important information about well performance and production forecasting. Using rate-normalized pressure (RNP), Samandarli et al. (2012) allow converting variable-rate and variable-pressure data to an equivalent of the drawdown pressure response to constant-rate production. This can allow recognition of flow regimes that can provide direct estimation of the formation permeability and the extent and efficiency of productive fracture in the estimation of stimulated reservoir volume, (SRV), Fig. 10.3. This will allow using observed field rates and pressures for matching with a global model to improve the estimates from the flow regime analysis. This methodology can be used to diagnose the condition of wells producing from unconventional resources.

With fracturing of horizontal wells, transient data are dominated by the behavior of the completion geometry, and traditional buildup data is impractical because long shut-in is required to examine the reservoir. The long-term production rate and wellhead flowing pressures can be used for reservoir evaluation. For deconvolution algorithms, buildup responses are needed that require more frequent tests than once a day using a wellhead pressure gauge. Linear flow is often seen during the transient period in unconventional gas wells that can last for several years. In unconventional oil and gas wells, the production rate is variable, and in many cases the extraction process is a constant pressure operation.

To analyze variable rate constant pressure data using pressure normalized rate method, a plot of 1/q vs. time (if reservoir pressure is constant) on a log-log plot can be diagnostic. Song and Ehlig-Economides (2011) also discussed the benefits of using long term data processed as rate-normalized pressure (RNP) and its derivative (RNP'). This approach uses the production data, which includes the majority of data acquired for a well, rather than using buildup data, which is a small fraction of the well data. More frequent buildup data can help in characterization of early time transient responses that can map the well completion geometry and/or near well reservoir heterogeneity.

As bottomhole pressure gauges are not normally installed in unconventional wells, the first step is converting wellhead flowing pressures to bottomhole conditions.

Decline in well performance can be a result of extremely low formation permeability, insufficient fracturing and low fracture conductivity that may be caused by crushing or embedment of propping agent Remedial actions may include use of more proppants for future wells. But it is often difficult to determine the cause of poor well performance.

As indicated by Brown et al. (2010), hydraulic fracturing can also alter the stresses in the fracture drainage area that may result in the formation of natural fractures in the near-vicinity of the horizontal well resulting in the characteristics of a trilinear flow model, representing linear flows in outer reservoir beyond the tips of the hydraulic fractures and the inner reservoir between hydraulic fractures and the hydraulic fracture.

This situation can be remedied in subsequent wells by creating longer fractures using larger treatment volumes, increasing the gel concentration perhaps using a smaller mesh proppant. If the cause is low fracture conductivity, in subsequent wells fracture treatments can include using a higher sand concentration or a stronger proppant.

The contribution of the reservoir beyond the stimulated volume can be negligible. For hydraulically fractured horizontal wells in unconventional reservoirs with micro-Darcy matrix permeabilities, a trilinear-flow model may represent the characteristics of flow toward a multiple-fractured horizontal well.

10.1 Transient Rate Analysis for Unconventional Reservoirs

In the analysis of performance data by the inverse rate plot, after a period signifying fracture dominated flow, there may be interchanges between the fractures. As indicated by Liu and Ershaghi (2021) and Linderman (2018), these can be noted with the observation of a slope of one developing after the fracture dominated follows on a log-Log plot of inverse rate vs. time. One cause can be considered cross flow among fractures. According to Liu and Ershaghi (2021), it can happen when some fractures stages such as B shown in Fig. 10.4 contribute more to the well flow and deplete faster

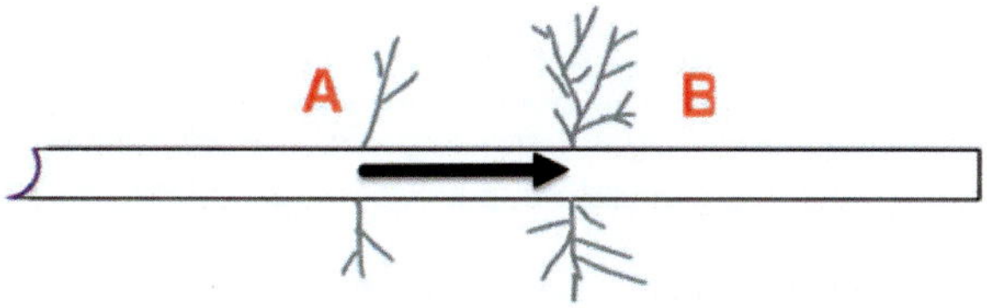

Fig. 10.4 Communication among the unequally pressured fractures

and experience a low pressure that may result in flow from high pressure fractured intervals (A) within the wellbore.

10.2 Questions and Answers

Q10.1: How can we recognize fracture cross flows in multi-stage fracturing in unconventional reservoirs?

A10.1: Fractured cross flow may be detected if a slope of one develops in the late time in a log-log plot of dp/q versus time.

Q10.2: How can we recognize the effectiveness of multistage fracturing from rate measurements?

A10.2: A long duration of fracture dominated flow observed on log-log plot of RNP (rate normalized pressure) vs. time can be indicative of the effectiveness of fracturing.

Q10.3: How can we detect cross flow among fractures?

A10.3: The formation of a unit slope at late times on a log-log plot of RNP vs. time may be an indicative of cross flow among fracture stages.

Q10.4: How can we detect boundary effects?

A10.4: Rapid increase of the slope on the log-log plot of RNP versus time can be caused by boundary effects.

References

Brown, M., Ozkan; E., Raghavan, R., Kazemi, H.: Practical solutions for pressure transient responses of fractured horizontal wells in unconventional reservoirs. SPE-125043-MS (2010). https://doi.org/10.2118/125043-M, https://doi.org/10.2118/0210-0063-JPT

Linderman, D.: Aspects of well productivity deterioration for unconventional wells. In: Paper presented at the SPE Annual Technical Conference and Exhibition, 24–26 Sept 2018, SPE-194029 (2018). https://doi.org.libproxy2.usc.edu/10.2118/194029-STU

Liu, Y., Ershaghi, I.: Impact of crossflow among fractures in multiply fractured horizontal wells in unconventional reservoirs. In: Paper presented at the SPE/AAPG/SEG Asia Pacific Unconventional Resources Technology Conference, Virtual, November 2021 (2021). https://doi.org.libproxy2.usc.edu/10.15530/AP-URTEC-2021-208363

Samandarli, O., Valbuena; E., Ehlig-Economides, C.: Production data analysis in unconventional reservoirs with rate-normalized pressure (RNP): theory, methodology, and applications, SPE-155614-MS (2012). https://doi.org/10.2118/155614-MS-2012.

Song, B., Ehlig-Economides, C: Rate-normalized pressure analysis for determination of shale gas well performance society of petroleum engineers (SPE). In: Paper presented at the North American Unconventional Gas Conference and Exhibition, 14–16 June 2011, Paper Number: SPE-144031-MS (2011). https://doi.org/10.2118/144031-MS

Chapter 11
Injection Well Tests

An important application of pressure transient test is fall-off testing. In studying water injection wells, besides the estimation of reservoir pressure around a given injector, study of injectivity with changing skin can be diagnosed. Additional estimations can be made with respect to the fluid interfaces formed. Well tests also provide an estimation of flow capacity and skin factor and can help in reservoir characterization.

Water injection may cause fracturing conditions that on the positive side may require fewer injection wells. But unplanned fracturing can also lead to the development of high permeability fractures that can negatively impact the areal sweep efficiency. Noak and Lester (1955) discussed some cases related to water injection fall off tests.

Injection well testing is a technique used in the oil and gas industry to evaluate the performance and characteristics of wells that are used for fluid injection into a reservoir. These tests provide important information about the injectivity and efficiency of the injection process, as well as reservoir properties and behavior.

During an injection well test, fluid (typically water, gas, or chemicals) is injected into the reservoir at a controlled rate via injection wells. The pressure and flow rate of the injected fluid, as well as the resulting pressure response in the wellbore and surrounding reservoir, are monitored and recorded over a specific period of time. The main objective of injection well testing is determining the injectivity of the well, which is a measure of how easily fluid can be injected into the reservoir. This helps in evaluating the performance of the injection system and identifying any potential issues or limitations. The data will also help to obtain reservoir parameters, such as permeability, porosity, and reservoir pressure, which are crucial for reservoir modeling and understanding the behavior of the injection fluid.

Interpretation of injection well test data involves analytical methods and numerical modeling techniques to analyze the pressure response, to estimate formation properties, and assess the effectiveness of the injection process. The information obtained from these tests helps in improving reservoir management, optimizing injection operations, and maximizing hydrocarbon recovery.

I. Ershaghi, *Solved Problems in Well Testing*,
https://doi.org/10.1007/978-3-031-47299-2_11

Pressure fall-off tests, also known as shut-in or build-up tests, are common methods used in the oil and gas industry to evaluate the properties and behavior of reservoirs. These tests involve shutting in a well after a period of production and monitoring the decline in pressure over time.

During a pressure fall-off test, the well is closed off from production or injection, and the pressure within the wellbore and surrounding reservoir is measured at regular intervals. The pressure decline that occurs during this shut-in period provides valuable information about the reservoir characteristics, such as permeability, skin damage, reservoir boundaries, and wellbore storage.

The main objectives of pressure fall-off tests include analyzing the pressure decline data, to determine the reservoir pressure at the well location. This information is essential foreservoir management and forecasting production performance.

We can further find important reservoir properties, such as permeability and skin damage. This information helps in understanding the fluid flow characteristics within the reservoir.

Pressure fall-off tests can provide insights into the extent and boundaries of the reservoir by examining pressure response and identifying boundaries where the pressure behaves differently. The data collected during a pressure fall-off test can be used to assess the efficiency of the well completion, identify any formation damage or interference, and optimize production strategies. Interpretation of pressure fall-off test data involves various analysis techniques, including type curve matching, analytical solutions, and numerical simulations, to analyze the pressure response and estimate reservoir parameters accurately. Overall, pressure fall-off tests are valuable tools in reservoir characterization and management, providing insights into reservoir properties and guiding decision-making processes related to well performance and production optimization.

Fall-off tests are associated with waterflood, steam flood or other injection projects. The principle of injection testing is applicable to every kind of lithological environment and reservoir fluids. As for conventional well testing, the presence of heterogeneities in multilayered or fractured reservoirs may complicate the interpretation. With the use of a production logging tool during the injection, one can determine the injection profile when testing a multilayer system and thus can reduce the interpretation uncertainties. In certain cases, because of the gravity-segregation effects and the difference in the density and compressibility of the fluids, injecting liquids into a gas reservoir complicates the interpretation of the system. Injection tests could provide all the information needed to calculate the well productivity at low costs and with a reasonable degree of reliability. Injectivity tests could also be applied successfully to a real sour-oil field.

In fall-off testing, brine injection could be a suitable solution as long as there are no compatibility problems with the oil, water and the reservoir rock. The interpretation of the pressure transients should be referred to the fall-off period rather than to the injection phase. The formation permeability-thickness product (k.h) could be identified correctly from the pressure-derivative analysis only if multiphase flow were assumed. The total skin value could also be obtained from the test interpretation. The total skin comprises of two components: a mechanical component resulting from permeability damage and a component related to fluid interaction in the reservoir. The fluid interaction related skin can be evaluated only with numerical well testing, given a reliable estimate of multi-phase relative permeability data.

The well head pressure does not disappear immediately when the injection well is shut in, and it is maintained by the expansion of the compressible fluids and by the elasticity of the formation. The residual pressure declines rapidly first and then slowly, to reach the static reservoir pressure.

11.1 Steam Injection

In recovering heavy oil, steam injection has successfully been used in many fields. The cumulative volume of steam injected can provide an estimate of the heat losses from the heated zone and affects heat efficiencies and the feasibility of the displacement of heavy oil. Expensive methods, such as coring and temperature observation wells during the passage of the displacement front, can be used. But to reduce the cost, well testing can be applied as a relatively quick and inexpensive method to obtain an estimate of the swept volume.

The thermal well testing method in steam flooding projects is based on pressure fall-off testing. This also applies to a composite reservoir model with two regions having highly contrasting fluid mobilities. The boundary between the inner and outer regions acts as a no-flow boundary. Consequently, the computed pressure response exhibits pseudo-steady state behavior ($dp/dt = \text{constant}$). Different methods can be used for the estimation of swept volume from pressure falloff tests including the deviation time, intersection time, type curve matching, and pseudo-steady-state methods.

During practical steam injection fall-off tests, the shut-in time is much less than the injection time. Thus, the MDH method of analyzing buildup (or fall-off) data can be used. When average steam properties are evaluated, liquid well testing analysis could be applied to steam fall-off testing. During the early time period of well tests, and after the end of short time wellbore storage effect, the infinite-acting radial flow occurs. The plot of pressure versus shut-in time yields a semi log straight line related to the flow capacity of the swept region. Using the slope of this semi log straight line, the steam effective permeability and skin factor may be calculated.

To select the correct straight lines, log-log pressure derivative function is used to identify various flow regimes. The semi-log pressure derivatives are calculated from the fall-off data using the differentiation algorithm. A log-log plot of the pressure derivatives function ($de_{ws}/d\ln(\Delta t)$) versus shut-in times can show a unit slope line for wellbore storage dominated flow, a constant derivative value for infinite-acting radial flow, and a unit slope line for pseudo steady state flow. Ziegler (1990) discussed a case study related to fall-of testing in a steam injection project.

11.2 Gas Injection

In pressure maintenance by injecting dry gas and cycling condensates, well testing can provide important information about the success of operations.

The injection of dry gas in rich gas/condensate reservoirs with pressure below the dewpoint pressure can be studied with well-test data to analyze vaporization of the condensate bank upon pressurization by gas injection. Because of significant differences in fluid properties and saturation distributions, well testing can help to evaluate the bank storage which then is used to calculate the bank radius. A problem in reservoir development is the formation of condensate deposit at pressures below the dewpoint pressure and the creation of a condensate bank that decreases productivity and may result in a potentially significant loss of liquid reserves. It is important to monitor and understand the behavior of the two-phase condensate bank during production when assessing its effect on productivity and ultimate recovery. Gas injection into rich gas/ condensate reservoirs attempt to maintain the average reservoir pressure and maintain the flowing bottomhole pressure above the dewpoint pressure. Pressure derivative function plots can be used to monitor changing fluid saturations in these reservoirs. Gas injection can help in re-vaporization under depressurization. Increasing pressure causes a change in fluid properties for both oil and gas and radially away from the well.

11.3 CO_2 Injection

Storage of CO_2 in subterranean geological formations including saline aquifers, depleted oil and gas reservoirs, coal seam gas reservoirs and shales can be a solution to the CO_2 sequestaration. Main concern is the cost of CO_2 capture affecting the development of large-scale CO_2 capture and storage projects.

Subsurface mature hydrocarbon bearing zones can be considered as primary targets for CO_2 sequestration where the operation cost can be offset by enhancing oil recovery and using the existing facilities. A subsurface formations with large volumetric void capacity is not necessarily a candidate for CO_2 storage as injectivity also needs to be examined for the selection of CO_2 storage. Well testing can help to examine the feasibility of injecting CO_2 in a well and examining the potential pressure leaks at monitoring wells.

11.4 Question and Answers

Q11.1: How can we recognize whether or not an injection operation has fractured the well?

A11.1: A fall-off test can help to identify the flow regime. If the well is fractured, a linear flow is manifested.

Q11.2: How can we interpret step rate tests in soft formations?

A11.2: In soft formations, increases in injection rates observed may relate more to increased injectivity caused by washout and removal of cementing materials holding grains than by fracturing.

Q11.3: How can we estimate reservoir pressures from fall off tests?

A11.3: The equilibrium pressures observed at late times in indicative of average reservoir pressures.

References

Nowak, T.J., Lester L.: Analysis of pressure fall-off curves obtained in water injection wells to determine injective capacity and formation damage. Trans. AIME **204**(01), 96–102 (1955). https://doi.org/10.2118/450-G

Paper Number: SPE-18140-PA: https://doi.org.libproxy2.usc.edu/10.2118/18140-PA

Ziegler, V.M.: Injection-well testing in a light-oil steam flood, Buena Vista Hills Field, California SPE. Prod. Eng. **5**(04), 394–402 (1990)

Chapter 12
Drill Stem Tests

DST stands for Drill Stem Test, which is a commonly used method in the oil and gas industry to evaluate the productivity and characteristics of a reservoir. During DST, a temporary packer or straddle packer is set in the wellbore, isolating a specific section of the reservoir. Once the packer is in place, a series of flow and pressure measurements are conducted to obtain important data about the reservoir. This can include flow rates, pressures, and fluid samples. The tool string typically consists of a bottomhole assembly (BHA) that allows fluid sampling, pressure measurement, and flow rate control. Pressure gauges are installed on the DST tool string to measure downhole pressures at different depths within the well. This data helps assess reservoir pressure, fluid gradients, and pressure boundaries. The tool string includes a downhole tester valve that allows fluids from the reservoir to flow into the tool string. This allows sampling and characterization of the formation fluids to assess reservoir properties including fluid composition, permeability, and productivity.

Drill Stem Testing provides an opportunity to check the condition and productivity potential of a formation exposed by drilling a well. We can examine the pressure, type of fluid and damage caused during the drilling (DST) is considered an important reservoir evaluation method for early evaluation of hydrocarbon reservoirs. Usually, several exploratory wells are drilled to evaluate the best exploitation scenarios, the number of production intervals, and their productivity potential. This is accomplished with a DST that consists of a pressure test using a temporary string that evaluates all potential intervals that were identified on open hole logs (Fig. 12.1).

Initial information from seismic, perforation data, open hole logs and DSTs are necessary for a thorough calculation of the potential reserves. DST is particularly important as it can obtain reservoir fluid samples at surface and bottomhole as well, measure hydrocarbon rates, and reservoir border effects. DSTs are also initiated after the well is drilled to evaluate hydrocarbon reserves at deep and ultra-deep seabeds. In deep and ultra-deep water prolonged exposure to low temperatures and heat loss can affect viscosity and may result in hydrate formation with the presence of gas.

I. Ershaghi, *Solved Problems in Well Testing*,
https://doi.org/10.1007/978-3-031-47299-2_12

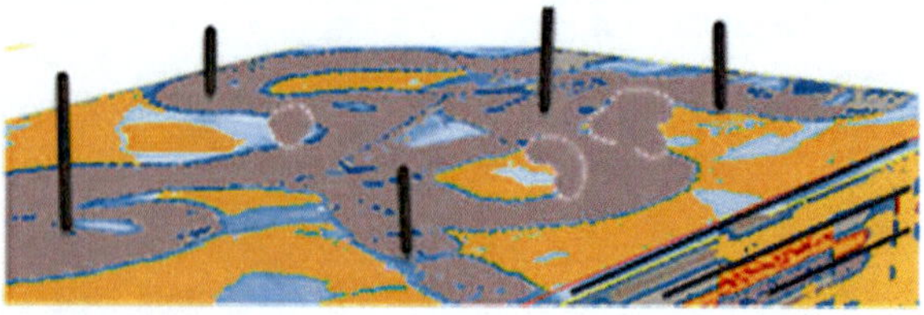

Fig. 12.1 Typical regions of a reservoir subject to DST testing

DST design besides the safety and risk mitigation must include analysis of hydrate formation and real-time data acquisition and real time decision making.

With a DST, both the flow and shuts-conditions are analyzed perform in build-up tests. Control of subsurface valves is with annulus pressure. Flow rates of hydrocarbon and water are measured using surface well-testing equipment. From a DST we may identify effective permeabilities to various reservoir fluids, Initial reservoir pressure, and depending on the duration of the test, boundaries, presence of heterogeneity, fluid contacts, skin affecting wellbore, fracture length and fracture conductivity if the well is fractured. DST's can be designed to obtain the flow behavior of the wells under actual production conditions. Information obtained from DSTs beyond permeability, skin, heterogeneity, and boundary information, includes well productivity for various flowing wellhead pressure conditions for field development decisions.

This information helps operators make informed decisions about completion designs, stimulation techniques, and production optimization strategies. Fluid samples obtained allow for analysis of the composition, quality, and characteristics of reservoir fluids. This can help in determining the type of fluid present, their phase behavior, and any contaminants that may be present. DST can indicate the presence of faults, compartmentalization, or communication with nearby wells, which influences reservoir management decisions. Interpretation of DST data consists of analyzing pressure response data and integrating the results with other well and reservoir data. This helps in estimating reservoir parameters, optimizing well and production strategies, and improving understanding of reservoir.

Behavior.

Earlier wells drilled used gauges that did not have the sensitivity of gauges today. Yet much useful information could still be inferred from such tests. For many decades drill stem testing (DST) has been used to determine reservoir conditions of potential producing zones, The early DST applications were limited to open holes at low to moderate temperatures and pressures. With the expansion of offshore DST to cased-hole applications with increased temperature ranges of above 400 °F and increased hydrostatic pressures exceeding 20,000 psi, traditional equipment experienced operational problems. These situations can result in deteriorating mud conditions caused by increased depth and temperature ranges, high pressures, and reduced sealing capability resulting from the high temperatures.

During a DST, borehole muds are subjected to mud deterioration because of mud remaining static at temperature for several days, causing the string to become stuck in the hole causing the formation of plugs of solids that prevents pressure transmission to the tools, or in oil-based muds that can cause an increase in viscosity and gel strength, affecting the transmission of operating pressures to the downhole tools.

A good DST chart must show a straight pressure base line, equal initial and final hydrostatic mud pressures flow and smooth buildup pressures. Wellbore conditions and malfunction of DST tool can be identified from the DST charts.

Open Hole Drill Stem Tests are temporary completion on a new well under drilling. The purpose is to test the productivity of a horizon, the type of fluid and reservoir pressure. Cased Hole DST are conducted after a well has been cased and completed. Hydrocarbon shows from the drilling mud, mud-logs, drill-cuttings, and coring can all help with the selection of intervals targeted for a DTS. Sometimes straddle packers or two cement plugs are used to isolate individual intervals for testing. In modern Drill Stem Testing, tools may include two or more pressure gauges, several packers, and a set of flow valves. The sequences of operations most importantly must consider the safety of the well using high density mud exerting pressures higher than the formation pressure to prevent formation flow.

By setting a packer above the formation, we can eliminate the effect of the mud column pressure. Packers are situated in the annulus between the test string and the casing or the well bore. To prevent any fluid entry into the string, a tester valve assembly is attached to the drill string and is positioned above the packer. The test string will also include pressure recorders as well as sampling chambers. The flow is initiated by opening the tester valve and energy from the formation pressure. The formation fluids can flow into drill string and then to the surface. The test consists of flowing and shut-in pressure tests. Formation fluid samples are also collected at the surface. If the fluid does not reach up to surface, it is collected from inside the test string. Several analyses can be done on formation fluid samples, gathered in the sample chambers.

Flow duration in open hole tests maybe short but during the cased hole testing the duration may be longer. If the formation is a gas producer, high surface pressures are observed with little or no liquid recovery from the test string. If the formation evaluated is oil bearing, producing surface pressures are observed and liquid recovered is oil usually free of water.

DST's can provide important information about various subsurface intervals. With advanced down-hole tools more robust data acquisition are possible for longer durations specially in offshore drilling, using floating rigs Introduction of coiled tubing have made the DST tests more effective, allowing down-hole shut-in, down-hole sampling, and placing or recovering pressure gauges during testing.

Using tubing conveyed perforation guns, rig time can be reduced, and perforation can develop under-balanced to minimize invasion of well bore fluid into formation. Besides recording of temperature and pressure from electronic gauges, pressure responses can be transmitted using a conductor cable. The simplest way to check the data quality is making sure the two recordings that relate to the static weight of the mud columns in the annulus are similar at the start of the test and at the end.

Because of the changing needs, we see improvements in DST technology. Such as open hole evaluation. Open hole evaluation has been associated with high risks because of test-packer leaks, the potential sand production and water breakthrough,

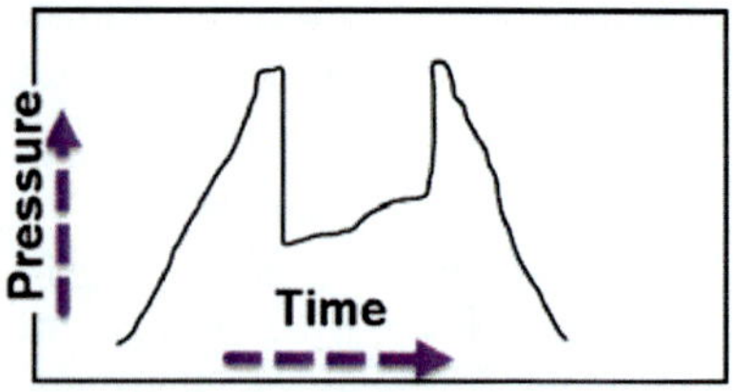

Fig. 12.2 Typical trend of DST pressure results

uncertainty about production and even cross flow. New developments allow setting the packer in the last casing and pinpointing the best intervals for evaluation from the knowledge of formation geology, minimizing the open whole sections. Using tools now available that include a new generation of pressure gauges for extreme reservoir conditions.

For unconsolidated formations, a sand-control completion can be implemented before running the DST string or by minimizing the drawdown during the flow period, to prevent destabilization of reservoir conditions. For low pressure reservoirs artificial lift may be needed during the DST. With the help of coiled tubing (CT) unit, it can include gas-lift pump, (ESP) system, jet pumps or a (PCP).

One of the most important factors to consider is the temperature profile at static conditions in the well, Low temperatures in deep water can have an impact on drilling, cementing, testing, completions, and stimulation. The cooling effect for the fluids from the reservoir is an important consideration. Besides the effect on viscosity, it can generate hydrates into the tubing.

When the bottom subsea tree is at the depth of the seabed, there is a need to use subsea safety system with the DST string. It will allow closing the subsea tree rams assuring the safety of the well at the seabed.

For hydrates formation, the conditions include high pressure, methane and water presence and low temperature. The formation of hydrates can plug the tubing. For a DST we need to consider methods that can optimize the test time required and reduce operational costs.

One needs to examine the data quality and stability of the gauges. A typical pressure recording during a DST looks as follows (Fig. 12.2).

DST's can also help in calibrating some petrophysical measurements.

Like other tests, the quality of recorded pressure data is important, and a diagnostic examination of the records can identify the flow capability of a formations. Below are some example responses when the two indicted data point are similar and or are dissimilar (Fig. 12.3).

The lack of agreement may not only relate to a gauge accuracy issue, but it may also be affected by the changes in the average density of the mud column. If some gas gets into the mud column, it can manifest itself by a lower mud column pressure. A leaking packer may also result in gas from the formation being sent to the mud column reducing the average density.

Among other pattern recognitions features are the recorded data related to the start and end of the flow. One expects that with the increase in the fluid height above and

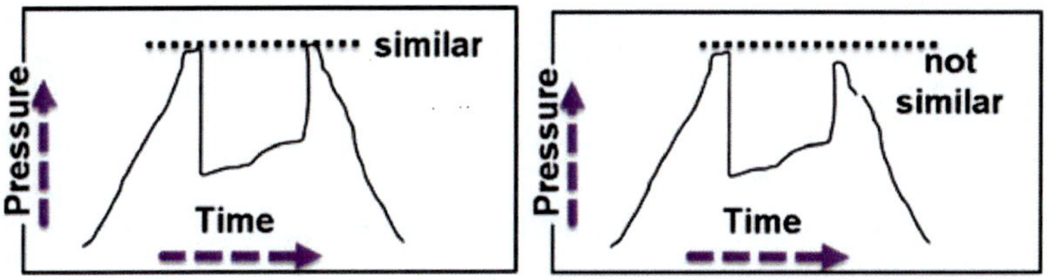

Fig. 12.3 Similar mud hydrostatic pressure points

Fig. 12.4 Example of pressure recording from a DST

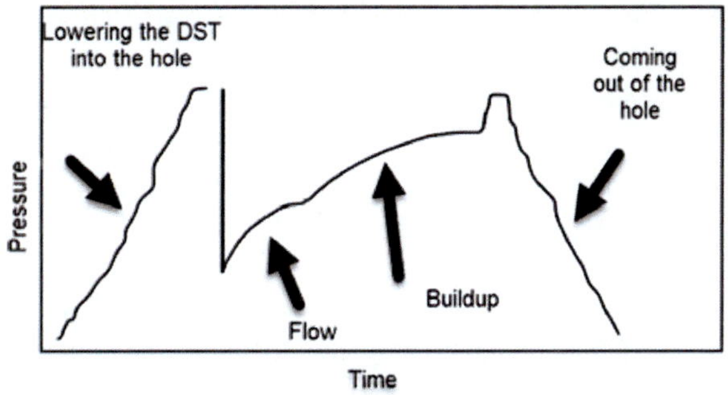

beyond the cushion fluid, the pressure at the end of the flow periods be larger than the values recorded at the start of the flow. When there is no change, one may interpret this as a sign that nothing has entered the drill pipe, indicating a tight formation. What is important is to also look at the buildup behavior right after each flow period.

The initial flow period is usually short to assure minimal pressure depletion from the test interval. There are times that the interval can be blocked by severe damage from the drilling fluid and no flow may be expected during the flow period. A good example is testing intervals that are heavily fractured. As an example, this was experienced in the drill stem testing of wells producing from the Monterey formation in offshore California. The DST response before and after acidizing indicated the effect of acids on fracture cleanups. It has become a routine to open up the fractures from the drilling mud residues before running a DST. DST tests can provide skin values that could have many causes.

Skin data can help to identify vertical communication, fracture plugging and other causes, geomechanical (stress dependent porosity/permeability induced) skin factor.

Once the quality of recorded data has been accepted, the flow periods can be analyzed by techniques presented by Ramey et al. (1975) and the buildup tests can be analyzed by using a pressure derivative function plot and a Horner plot (Fig. 12.4).

In drill stem testing there is now interest in the use of wireless telemetry systems. These systems transmit real-time data from downhole tools to surface equipment, enabling operators to monitor and analyze data in real-time. These systems use electromechanical or electrochemical means to transmit data through the drill string, eliminating the need for traditional drill pipe-mounted sensors. An advantage of wireless telemetry systems is that they eliminate the need for cable-based systems, thereby reducing costs and improving safety. The systems also allow for faster data acquisition and analysis, allowing more informed decisions in real-time.

Also, the development of real-time downhole sensors, which can measure various parameters such as pressure, temperature, and flow rate can provide operators with more accurate and precise measurements, enabling them to optimize well production and improve recoveries.

Use of telemetry was dissuaded by Haddad et al. (2010). Other issues related to the offshore application of DST and high pressure and temperature condition have been discussed by Ringgenberg and Self (1996) and Salguero et al. (2011). Also in some cases, the method of closed chamber DST may be necessary, and the reader is referred to Alexander (1977) paper.

In summary, (DST) is an important reservoir evaluation method for early evaluation of subsurface reservoirs. Usually, several exploratory wells are drilled to evaluate the best exploitation scenarios, the number of production intervals, and their productivity potential. This is accomplished with DST 's that consist of a pressure test using a temporary string that evaluates all potential intervals that were identified on open hole log results. Initial information from seismic, perforation data, open hole logs and DSTs are necessary for a thorough calculation of the potential reserves. The DST is particularly important as it can obtain reservoir fluid samples at surface and bottomhole as well, measure hydrocarbon rates, and other reservoir border effects.

12.1 Questions and Answers

Q12.1: How can we estimate the presence of skin from DST?

A12.1: A Horner plot of the first buildup shows skin before the development of plot of P versus the log (Horner time). There may be lessening of skin on the second buildup if during the flow some of the materials build up causing the skin are removed by the production.

Q12.2: How can we estimate properties from the flow period of a DST?

A12.2: The flow period pressure data can be analyzed by a method discussed by Ramey et al. (1975).

Q12.3: How can we easily check the quality of the data recorded?

A12.3: Hydrostatic pressure corresponding to the mud column before the packer is set and after it is released should be similar.

Q12.4: How can we recognize no flow into the test equipment?

A12.4: There will be no increase in the recorded pressure corresponding to the onset of tool opening and flow period.

References

Alexander, L.G.: Theory and practice of the closed-chamber drill stem test method. JPT **1539**, 7 (1977)

Haddad, J., Salguero, A., Jaimes, C.: Application of telemetry technology in high-pressure wells to improve data accuracy in drill stem tests. SPE-129160-MS (2010). https://doi.org/10.2118/129160-MS

Ramey, H.J. Jr., Agarwal, R.G., Martin, I.: Analysis of 'slug test' or dst flow period data. J. Cdn. Pet. Tech. 1–11. **658**, 21 (1975)

Ringgenberg, P., Self, J.C.: Increased reliability for HP/HT drill stem testing from full scale tool tests. OTC-8191-MS (1996). https://doi.org/10.4043/8191-MS

Salguero A., Wendler, C., Mansilla; C., Woolsey, S.: Worldwide drill-stem-testing experiences in heavy and viscous-oil offshore environments that improve operational efficiency, SPE-150495-MS (2011). https://doi.org/10.2118/150495-MS-2011

Chapter 13
Computer Aided Methods and AI

Advances in gauges and the need for rapid interpretation of well-tested data have demonstrated a need for a computer-aided approach to interpretation. The well-test interpretation procedure can potentially be completely automated by implementing various approaches. In automated type-curve matching, the selection of an appropriate reservoir model and the initial parameter estimation are essential for obtaining reliable results.

In well test analysis, the selection of the interpretation model is the most difficult task. In manual well test interpretation, selection of the interpretation model usually requires human expertise in pattern recognition. In this book, we have used graphical tools to aid in the model selection process. These graphical techniques are simple and require minimal calculations.

When a plot of the log-log pressure derivative function vs. time is used, the recognition of the true reservoir response can be complicated. For human interpretation, this can sometimes be managed, but implementation with computer assisted interpretation can be difficult. Computer use, however, can significantly help in terms of data analytics and graphic display of data and for model selection.

We have discussed how, from a log-log plot of the pressure derivative function vs. time, one can identify reservoir models by recognizing features that are specific to particular analytical models. The selection of an interpretation model by a human expert is a visual process. Computers can expand the capabilities of well testing and this has been discussed by many. Computers can speed up all the specialized analyses.

Mattar (1996) discussed a multitude of effects other than the intended reservoir pressure transients including multiphase, geotidal, micro seismic, changing liquid levels, recorder drift, recorder plugging, that can mask the reservoir transient. He indicated that pressures must be corrected to remove the non-reservoir effects before pressure-transient analysis is attempted, otherwise, diagnosis of the observed data can be misleading. He also indicated that data preprocessing is a significant component of well-test interpretation.

I. Ershaghi, *Solved Problems in Well Testing*,
https://doi.org/10.1007/978-3-031-47299-2_13

Artificial intelligence (AI) methods can play a significant role in well test analysis. They offer the potential to enhance efficiency, accuracy, and decision-making throughout the process. AI algorithms can process large volumes of well test data, including pressure, flow rates, and fluid properties to identify patterns and relationships. By utilizing machine learning and data analytics techniques, AI can provide faster and more accurate analysis of well test data, helping to derive meaningful insights and identify reservoir parameters.

AI algorithms can also be trained to detect anomalies or deviations from expected behavior in well test data. This can help identify issues such as equipment malfunctions, data errors, or unexpected reservoir behavior. Early detection of anomalies enables timely corrective actions and improves the overall reliability of well test analysis.

AI can assist in optimizing the design of well tests by analyzing historical data and simulation models. By considering various parameters and constraints, AI algorithms can suggest optimal test durations, flow rates, test sequences, or packer positions to maximize the information obtained from the well test while minimizing costs and risks.

AI can be used to develop predictive models that continuously monitor and analyze real-time data during a well test. This allows for early detection of potential issues or variations from expected behavior, enabling proactive decision-making and adjustments during the test.

AI techniques can also facilitate the integration of well test data with reservoir models. By combining well test data with geological and reservoir simulation models, AI algorithms can help refine reservoir characterization, for estimating reservoir parameters, and improve predictive capabilities.

AI can also provide decision support tools for well test analysis by assessing uncertainties and quantifying risk. Through probabilistic modeling and scenario analysis, AI can help evaluate different interpretations and uncertainties associated with well test results, aiding in more informed decision-making.

13.1 Application of AI

Allain and Horne (1990) indicated that AI can help in model identification and parameter estimation. They suggested an algorithm that can distinguish response where there is noise in the data. One application of machine learning is programming computers to automatically recognize patterns in well-tested data. This is different from traditional programming systems where explicit rules are defined and hard coded into a computer program for doing a specific task. With a machine learning approach, we do not "teach" a computer how to do something; instead, we expose it to a well test dataset and let it learn how to do a task on its own based on patterns it found in the dataset.

Rosa and Horne (1983) studied the use of nonlinear regression. They developed techniques that simplified implementation of any reservoir model in an automated procedure. Allain and Horne (1990) used synthetic pattern recognition and a rule-based system to discern the reservoir model by extracting symbolic data from the pressure derivative data. The well and reservoir parameters were also estimated. The limitations were that the methods required a preprocessing of the derivative data to distinguish the true response from the noise and a defining rule to accommodate 'nonideal' behavior.

An early application of ANN in the oil industry was proposed by Juniardi and Ershaghi in 1993. Some have applied the improved backward propagation neural network to the forecasting of oilfield production. Now, ANN is an important branch of ML, and it has wide applications in the oil industry. ANN is an intelligent learning model used to mimic the structure and function of the neural network of the human brain.

Al-Kaabi and Lee (1990) used artificial neural networks (ANN) to identify well test interpretation model from pressure derivative data. They indicated that this approach was effective in identifying reservoir models without the need for data smoothening. Allain and Houźe (1992) presented a hybrid approach to combine symbolic and artificial neural network methods. They suggested the use of the neural network to determine the sketch of the derivative before applying a rule-based approach to determine the model and estimate its parameters. Ershaghi et al. (1993) implemented multiple neural networks with each neural network representing a single reservoir model. The purpose was to overcome the inefficiency in the training of an enormous number of reservoir models. Their proposed approach assists in recognition of various models with similar responses. In their approach, output from the neural network was cross correlated with models suggested from other data sources. They did not discuss parameter estimation, but their approach improved the shortcomings of the artificial neural network approach as a tool in well-test interpretation. Tian and Horne (2019) discussed different machine-learning techniques to characterize wellbore-storage effect, skin effect, flow model flow, and boundary effect. Anraku and Horne (1995) introduced a new approach to differentiate reservoir models using the sequential predictive probability method. Athichanagorn and Horne (1995) studied the use of the artificial neural network and the sequential predictive probability approach to identify characteristic components of potential models on the derivative plot. They showed that one may identify the flow regimes that corresponded to these characteristic components. Dastan and Horne (2011) made improvements in using non-linear regression in well test analysis by overcoming various issues related to noise.

In brief. AI in well test analysis has the potential to enhance efficiency, accuracy, and decision-making in reservoir evaluation, well performance assessment, and production optimization. It enables faster and more comprehensive analysis of well test data, leading to improved reservoir understanding and effective reservoir management strategies.

13.2 Need for De-Noising

Downhole gauges can generate large volumes of data at small time intervals. There can be errors associated with physical changes in a reservoir and this can further complicate the interpretation task. Noisy pressure data can result in wrong interpretations and uncertainties in estimated parameters. The errors can be systematic or can represent random deviations from the true values. There are also systematic errors caused by inaccuracies in the system that may relate to calibration or random errors in the reading of gauges. As such, there can be errors associated with noise. This can be caused by measurement errors. The more common type of noise is having random noises. Distinguishable from noise are outliers that can generate uncertainties in well test data. Computers can help in denoising the data before the interpretation.

A number of methods have been introduced such as the wavelet analysis. There can be several stages including outlier removal, denoising, transient identification, data reduction, flow rate history reconstruction, behavioral filtering, and data interpretation. Zhao and Reynolds (2009) introduced a method for removal of the tidal component of the bottomhole pressure change from the measured pressure.

13.3 Questions and Answers

Q13.1: How can we recognize noise in well test data?

A13.1: In pressure transient analysis of data, noise refers to responses to some physical phenomena not related to the reservoir, including segregation of fluids in the wellbore and pressure leaks, which can influence the data and the interpretation. Denoising algorithms can help to remove such anomalies and to extract the data that can help to identify the reservoir model.

Q13.2: What are the filtering methods to remove noise?

Q13.2: Several techniques have been suggested and the use of wavelet techniques seems to offer better results.

Q13.3: What is needed for machine image analysis using neural networks to identify the appropriate reservoir model?

A13.3: The system must be trained for various reservoir conditions first.

References

Al-Kaabi, A.U., Lee, W.J.: Using artificial neural networks to identify the well test interpretation model. In: Paper SPE 20332 presented at the 1990 5th SPE Petroleum Computer Conference, Denver, Colorado, 25–28 June (1990)

Allain, O.F., Horne, R.N.: Use of artificial intelligence in well test interpretation. J. Petroleum Tech. pp. 342–349 (1990)

Allain, O.F., Houźe O.P.: A practical artificial intelligence application in well test interpretation. In: SPE 24287 presented at the 1992 SPE European Petroleum Computer Conference, Stavanger, Norway, 25–27 May (1992)

Anraku, T., Horne, R.N.: Discrimination between reservoir models in well test analysis. In: SPE Formation Evaluation (June), pp. 114–121 (1995)

Athichanagorn, S., Horne, R.N.: Automated parameter estimation of well test data using artificial neural networks. In: SPE 30556, presented at the 70th Annual Technical Conference and Exhibition, Dallas, TX, 22–25 October (1995)

Dastan, A., Horne, R.N.: Robust well-test interpretation by using nonlinear regression with parameter and data transformation. SPE J. (2011). https://doi.org/10.2118/132467-PA

Ershaghi, I., Li, X., Hassibi, M.: A robust neural network model for pattern recognition of pressure transient test data. In: SPE 26427 presented at the 1993 SPE Annual Technical Conference and Exhibition, Houston, Texas, 3–6 October (1993)

Mattar, L.: Critical evaluation and processing of data before pressure-transient analysis. SPE Form Eval. **11**, 120–127 (1996). https://doi.org.libproxy1.usc.edu/10.2118/24729-PA

Rosa, A.J., Horne, R.N.: Automated type curve matching in well test analysis using Laplace space determination of parameter gradients. In: Paper SPE 12131, presented at the 1983 SPE Annual Technical Conference, San Francisco (1983)

Tian, C., Horne, R.: Applying machine-learning techniques to interpret flow-rate, pressure, and temperature data from permanent downhole gauges. SPE Res. Eval. & Eng. **22**(02), 386–401 (2019). https://doi.org/10.2118/174034-PA

Zhao, Y., Reynolds, A.C.: Estimation and removal of tidal effects from pressure data. SPE J. **14**, 144–152 (2009). https://doi.org.libproxy2.usc.edu/10.2118/103253-PA

Chapter 14
Pumping Wells

When a production well is on artificial lift, it can still be tested and depending on the methodology used to obtain the data, the information can be processed to gain insight into the nature of the conditions around the well and the formation.

Rod pumps are the major type of altricial lift methods. In general, the goal is to estimate the fluid level rise and associate that with the rise in formation pressure. The liquid height above the pump represents the pumpable liquid the pump can produce. Data can also be acquired from many types of rod-pump completions, including wells with downhole gas separators. This system allows more flexibility in well testing and allows for greater accuracy. Combining fluid level measurement with surface pressure data for estimation of the BHP (bottom hole pressure), introduce errors accounting for multiple fluid gradients in the wellbore. With no accounting for gas in solution during a buildup. Information from fluid-level may not address the high-resolutions necessary for derivative-type curve matching with modern well-test-analysis software.

Pressure transient testing of rod pump wells includes analyzing the pressure responses within the well and formation to obtain information about the well and reservoir performance. The primary objective of such testing is to estimate well parameters, reservoir properties, and evaluate the efficiency and productivity of the rod pump system. A shut-in buildup test on rod pump wells consists of shutting down the pump and allowing pressure to build up within the well and reservoir. By monitoring the pressure response over time, parameters such as skin factor, wellbore storage capacity, and near-wellbore properties can be estimated. These estimates help in understanding the reservoir's productivity and the well's efficiency. Pressure transient tests also provide insights into the performance of the rod pump wells. For example, these tests can identify the presence of damage or skin near the wellbore that may be restricting fluid flow. This information can guide remedial actions or optimization strategies to improve performance. Overall, pressure transient testing of rod pump wells can aid in estimating reservoir properties, diagnosing well issues,

I. Ershaghi, *Solved Problems in Well Testing*,
https://doi.org/10.1007/978-3-031-47299-2_14

optimizing production rates, and supporting effective reservoir management decisions. Case studies have been presented by Gyuoynes et al. (2000) and Brownscombe (1982).

With various completions such as gas anchors or downhole gas separators, electronic memory gauges can be run on the rod string with the downhole pump assembly, in both vertical and horizontal completions allowing downhole data measurements during both flow and shut-in conditions. This will allow high-resolution pressure data for a more accurate well-test analysis. This technique can be run when a packer is used in the completion to minimize wellbore storage effects and test time. The gauge can be run on the rod string without pulling the completion packers or tubulars. This is advantageous to using wireline-conveyed pressure gauges, tubing-conveyed pressure-recording devices, and surface pressure monitoring systems requiring fluid-level measurement to record data from pumping well completions.

The operator may run the gauge assembly below the pump seat if there is no downhole gas separator. or the gauge assembly can be set in between the rod pump and the pump seat. This eliminates the requirement that the tubing string be pulled out while also allowing the recording of the flowing downhole pressure.

Pressure transient analysis of ESP (Electric Submersible Pump) wells involves analyzing the pressure responses within the well and the reservoir to gather information about the well performance and reservoir characteristics. ESPs are commonly used in oil and gas wells to lift fluids to the surface. By examining the pressure response, one can identify issues like a malfunctioning pump, tubing or casing problems, scaling or fouling of the equipment, or other factors that can impact the pump efficiency and overall, well performance and reservoir development. Case studies are discussed by Camilleri et al. (2021).

Gas lift wells typically undergo an injection period followed by a buildup period. During the injection phase, gas is continuously injected into the well to lift the fluids to the surface. Later, the injection is stopped, and the pressure buildup response is monitored. By analyzing pressure data during the buildup period, important parameters can be estimated. Analyzing these relationships helps to optimize injection rates, detect any operational issues affecting well performance, and determine the best operating conditions for the gas lift systems. Some case studies are discussed by Jain and Ayoub (1984).

14.1 Questions and Answers

Q14.1: What is an important parameter that can be obtained from a pumping well?

A14.1: Condition around the well as affected by perforation plugging, buildup of skin and finally, the static reservoir pressure.

Q14.2: Could high-resolution downhole pressure gauges be installed to rod pump well?

A14.2: Yes, that can be done by using a carrier assembly.

Q14.3: Is the assumption of incompressible fluid in computation of fluid rise realistic?

A14.3: Well fluid especially if it contains any gas is compressible fluid and estimations using fluid level rise may introduce errors.

Q14.4: To reduce the cost and the operational risk, many gas lift wells are evaluated with surface shut-in and bottomhole pressure measurement. What are some of the concerns?

A14.4: Wellbore storage is a concern with a surface shut-in. Wellbore storage pressure may sometimes behave like radial flow. This factor needs to be considered,

Q14.5: Can a gas lift well be evaluated with conventional wireline memory gauges?

A14.5: Yes, pressure testing can be conducted with conventional downhole electric line unit.

Q14.6: What are some problems that can develop in testing of gas lift wells?

A14.6: The most important is the expansion of annulus gas that can cause the appearance wellbore storage.

Q14.7: How can the wells equipped with ESP be evaluated especially if a long-term test is desired?

A14.7: One can use permanent downhole pressure sensors. Sometimes traditional wireline gauges are also installed prior to conducting the tests.

Q14.8: What are some issues associated with the wells equipped with ESP that can affect the pressure transient data?

A14.8: In ESP wells, noise, production history, time-synchronization issues, and wellbore effects can impact the measured signals.

References

Brownscombe, E.R.: Afterflow and buildup interpretation on pumping wells. SPE 8354 (1982)

Camilleri, L., Al-Hussainee, N., Al-Janabi, M., Al-Joran, M., Mohammed Kamal Aal Najar, M., Ayoub, J.: Delivering pressure transient analysis during drawdown on ESP wells: case studies and lessons learned D031S024R007 (2021). https://doi.org/10.2118/204567-MS

Gyuoynes, J.C., Azari, M., Gilstrom, R., Friend, B.I., Firbanks, D.: New well testing methods for rod pumping oil wells—Case studies. SPE 63082 (2000)

Jain, A., Ayoub, J: Pressure buildup in gas lift wells. JPT (1984)

Chapter 15
Test Design

The goal of reservoir characterization is to understand and define the reservoir and its fluid properties for optimizing production. As indicated before, we use information from various sources. Among different tools and techniques, well tests are considered important reservoir characterization tool. They help to confirm the analysis and interpretation of data obtained on wells from other tools.

A very important aspect of a well test is that it is designed to fit the purpose. For a single well test, such as a buildup test, the objectives must be clear. For example, if the goal is to assess the gradual buildup of skin during the production operation, the test can have a short duration. If the goal is to look deeper into a reservoir, longer test times are needed. In this process, the transmissibility of the formation plays an important role. If the goal is to look deep and identify boundaries, larger test durations may be needed depending on the distance from the boundary to the well. Additionally, the delays caused by the presence of wellbore storage must be considered. Proper Well test designs can help in obtaining formation characteristics and well production behavior which will increase the value of information from well tests.

The following formula suggested by Earlougher (1977) is commonly used in the design of drawdown tests. The estimate provides minimum constant rate flow time to go pass the wellbore storage control is:

$$t = \frac{C * (200,000 + 12000 * S)}{\mathrm{T}} \tag{15.1}$$

where (c) is the wellbore storage in Bbl./psi and (S) is the dimensionless skin factor. As shown in Fig. 15.1, the required flow time to see pass the wellbore storage effect is substantially influenced by the formation transmissibility and wellbore storage.

For larger values of skin factor, the time requirement is substantially higher, Fig. 15.2.

I. Ershaghi, *Solved Problems in Well Testing*,
https://doi.org/10.1007/978-3-031-47299-2_15

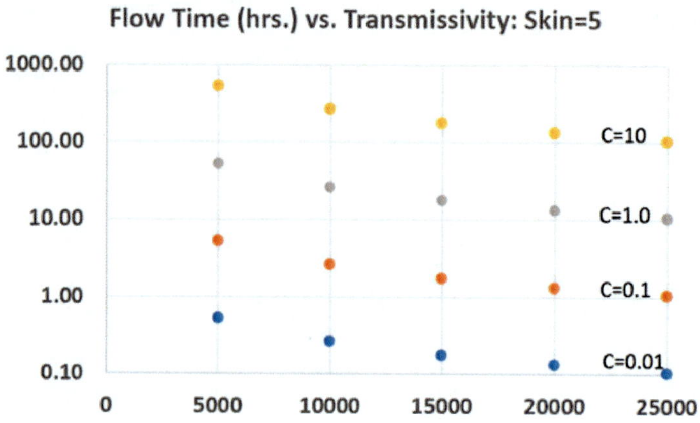

Fig. 15.1 Flow time and transmissibility relationship at S = 5

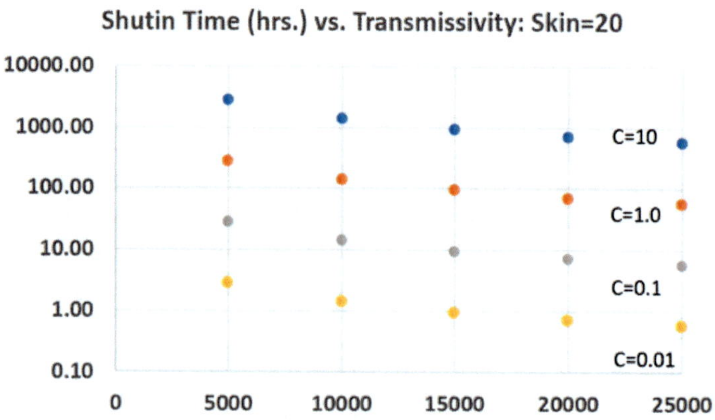

Fig. 15.2 Flow time and Transmissibility Relationship at S = 20

For the estimation of buildup time, we can use an approximation shown by Spivey and Lee (2013).

$$t = \frac{(647{,}000 + 37{,}700{*}S){*}C}{T} \tag{15.2}$$

For a reasonable shut-in time, there should be negligible wellbore storage effect for high skin factor conditions.

The test duration depends on formation transmissibility as well as formation porosity and total system compressibility. Based on the study published by Earlougher and Kazemi (1980), the practical time required to investigate a boundary is about the time it takes to investigate a distance of roughly four times the distance to the boundary.

$$t = \frac{Porsoity\,.\,Ct\,.\,viscosity}{0.00105\,.\,K}\,.\,d^2 \tag{15.3}$$

For example, for a boundary located about 200 ft., the shut-in time should be what it takes to investigate roughly 4 * 200 = 800 ft. For a reservoir permeability of 500 mD., porosity = 0.2 and Ct (compressibility) = 50e−6 1/psi, viscosity = 1 cp, the required time is:

$$t = 12\,\text{h} \tag{15.4}$$

For a tighter reservoir such as K = 50 mD., the required shut-in time will be 120 h.

15.1 Questions and Answers

Q15.1: What are the considerations for designing the duration of a sufficient buildup test?

A15.1: If the measurement is done at the wellhead, the most important consideration is that the test is run long enough to record formation response data beyond the wellbore storage period.

If the gauge is downhole or is of the permanently installed type, then the transmissibility of the formation may be estimated. For tight reservoirs it takes longer to investigate the reservoir.

Q15.2: How can we design a multiple well interference test?

A15.2: Distance between the wells and the range of the estimated transmissivities between the wells must be considered. To record a reasonable pressure response, the longest distance must be used to generate the highest reassure responses in all the wells during the test.

Q15.3: How do we determine the type of gauge needed for pulse and interference test?

A15.3: For pulse tests, we measure both the lag time and the pressure change and, as such, both the accuracy of the pressure gauges and high resolutions are important. In interference tests, we need high resolution gauges that can detect very mall pressure changes.

References

Earlougher, R.C. Jr., Kazemi, H.: Practicalities of detecting faults from buildup testing. J. Pel. Tech. 18–20 (1980)

Earlougher, R.C., Jr.: Advances in Well Test Analysis, vol 5. Monograph Series SPE (1977)
Spivey, J.P., Lee, W. J.: Applied well test interpretation. SPE Textbook Series, vol. 13, p. 334 (2013)

Chapter 16
Ground Water Hydrology

The use of well testing in groundwater hydrology helps in studying the behavior and characteristics of aquifers, and management of groundwater resources. This is done through conducting controlled pumping or injection tests in wells to obtain data that can be used for aquifer characterization and hydraulic parameter estimation. From well testing by using the water levels we can estimate hydraulic conductivity, transmissivity, storativity, and specific yield.

Analysis of water level responses during pumping can also allow for the calculation of the aquifer properties that can help in determining the yield, flow pathways, and recharge rates of the aquifer. Well testing can also allow for the assessment of well performance, including well efficiency, drawdown, and specific capacity. By monitoring the drop in water levels and the rate of recovery during and after pumping, we evaluate the pumping efficiency, the impact of pumping on neighboring wells, and the long-term sustainability of the well.

Such well tests can also help in estimating parameters related to groundwater flow including the hydraulic conductivity and storage coefficient. These parameters are important for developing numerical models and predicting groundwater flow patterns, for contaminant transport, and the effectiveness of remediation strategies. From the response of water levels to pumping or injection in different wells within an aquifer system, we can potentially infer the boundaries of the aquifer, identify potential barriers or confining layers, and assess the storage properties of the subsurface formations. This can help in delineating water availability, predicting well interference, and managing groundwater resources. From such well tests we can also obtain valuable data for calibrating numerical models that simulate groundwater flow. The observed pumping or injection data, combined with water level responses from monitoring wells, can help in refining model parameters and confirm model predictions, and improve the accuracy of groundwater flow simulations.

These tests further help in determining the optimal pumping rate for a well to meet the desired water supply requirements. By gradually increasing the pumping rate and monitoring the water level response, we can identify the sustainable pumping rate

I. Ershaghi, *Solved Problems in Well Testing*,
https://doi.org/10.1007/978-3-031-47299-2_16

that balances the well's capacity with the aquifer's recharge capabilities. This is done by measuring the drawdown caused by pumping. The drawdown can indicate the hydraulic response of the aquifer to pumping and can be used to estimate important parameters like the aquifer's transmissivity (T) and storativity (S). These well tests can provide insights into the well's storage properties, hydraulic connectivity with adjacent aquifer zones, and the presence of potential barriers or confining layers.

The analysis of the pumping test data also helps in understanding the presence of different hydrogeological units, and the potential for well interference. Such test data is also important for calibrating groundwater flow models. The observed pumping rates, drawdowns, and recovery data provide valuable information to refine model parameters and to validate the model's ability to simulate pumping well behavior and its impact on the aquifer.

A simple solution that can be used is the Theis equation. The Theis (1935) equation is for flow to a fully penetrating line sink producing at a constant rate in a homogeneous, isotropic and nonleaky confined aquifer of infinite extent is as follows:

$$\mathrm{s} = \frac{Q}{4\pi T} \int_{u}^{\infty} \frac{e^{-y}}{y} \mathrm{dy} \tag{16.1}$$

$$\mathrm{u} = \frac{r^2}{4T.t} \mathrm{S} \tag{16.2}$$

The integral is called the *Theis well function.*
Where:

$$\mathrm{S} = \frac{Q}{4\pi T} w(u) \tag{16.3}$$

$$\mathrm{W(u)} = -0.5772 - \ln(\mathrm{u}) + \mathrm{u} - \frac{U^2}{2.2!} + \frac{U^3}{3.3!} - \frac{U^4}{4.4!} + \cdots \tag{16.4}$$

For partially penetrating pumping well, Hantush (1961) extended the Theis method to include partial penetration effects for nonleaky confined aquifers. Mueller and Witherspoon (1965) expanded the application to interference test among wells in aquifers.

In computing T and S, we use the groundwater hydrologist units, T (Transmissivity m^2/s) and S will be dimensionless.

16.1 Questions and Answers

Q16.1: Find the T and S for an aquifer given the following water level data producing at Q = 3 m^3/s and located 50 m from the monitoring well (Fig. 16.1).

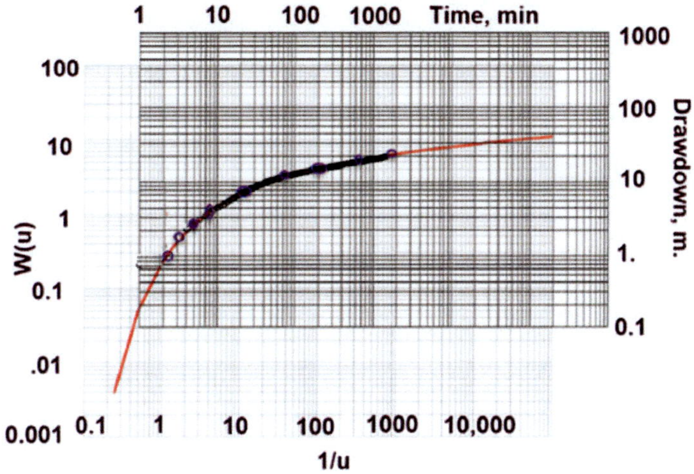

Fig. 16.1 Matching the water level data on a Theis curve

t (min)	S (m)	t (min)	S (m)
2	1	20	8
3	1.75	70	13
4.5	2.5	700	18
7	4	1800	20

A16.1: By plotting the points on a Theis curve we estimate T and S.

At a match point: 1/u = 100 W(u) = 4 t = 200 min Drawdown = 15 m

$$\mathrm{T} = \frac{\mathrm{Q} * \mathrm{w(u)}}{4\,\mathrm{S}\pi} = \frac{3 * 4}{4 * 3.14 * 15} = 0.0636\,\mathrm{m}^2/\mathrm{s}$$

$$\mathrm{u} = \frac{r^2}{4T.t}\mathrm{S}$$

$$\mathrm{S} = \frac{4 * 0.0636 * 200 * 60}{(50 * 50)} = 0.0122$$

References

Hantush, M.S.: Drawdown around a partially penetrating well. J. Hyd. Div. Proc. Am. Soc. Civil Eng. **87**(HY4), 83–98 (1961).

Mueller, T.D., Witherspoon, P.A.: Pressure interference tests within reservoirs and aquifer's. JPT (1965)

Theis, C.V.: The relation between the lowering of the piezometric surface and the rate and duration of discharge of a well using groundwater storage. Am. Geophys. Union Trans. **16**, 519–524 (1935)

Chapter 17
Appendix

17.1 Problems and Solutions

These sample problems are mixed and include different topics. Readers are invited to solve the problems, review the solutions, and, if needed, refer to the corresponding sections discussed in the book.

Q1-Complete the table:

Question	I recommend the following tests:
Distance to a fault	
There is a fault between two wells:	
The effective length of a fracture in a tight Unconventional reservoir:	
AOF	

A—Pressure Buildup
B—Pressure Drawdown
C—Back Pressure Test
D—Pulse test
E—dp/q versus Time

Answer: A, D, E, C.

Q2-The following test data was obtained. Interpret this derivative function plot.

I. Ershaghi, *Solved Problems in Well Testing*,
https://doi.org/10.1007/978-3-031-47299-2_17

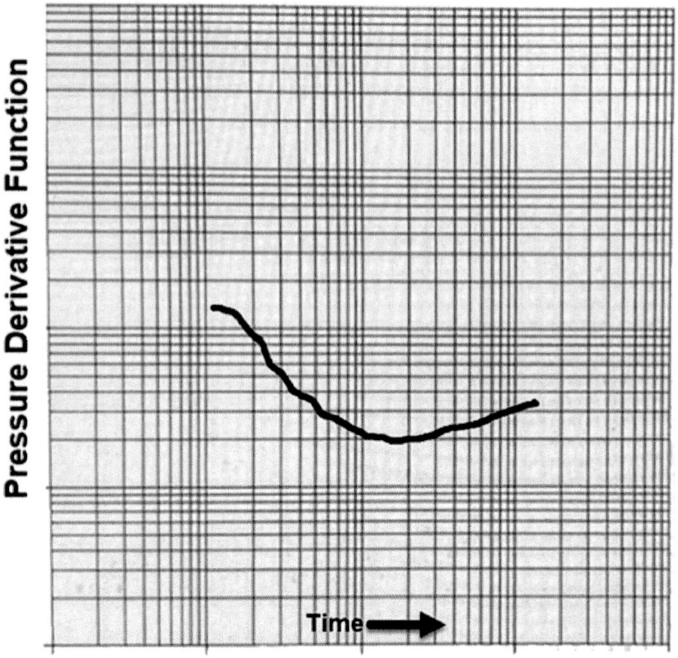

Answer:

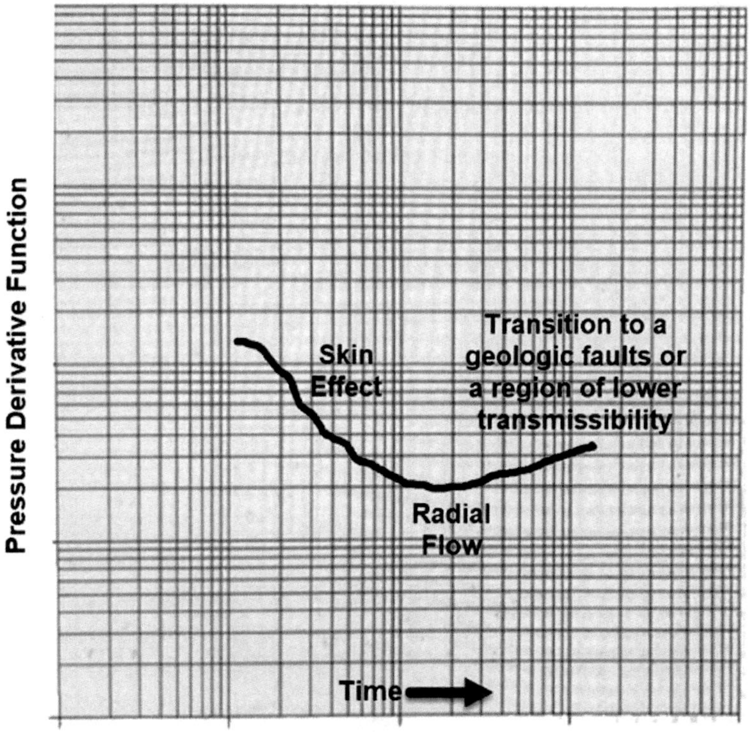

Q3-Given the following pressure derivation plot, interpret and explain each segment.

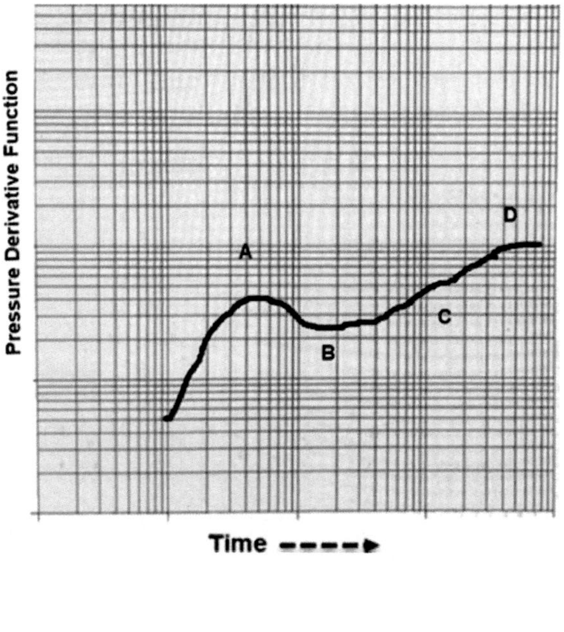

Answer: A: Wellbore storage and skin B: Radial flow C: Linear flow D: Radial flow.

Q4-Make two interpretations and explain your reasons for each:

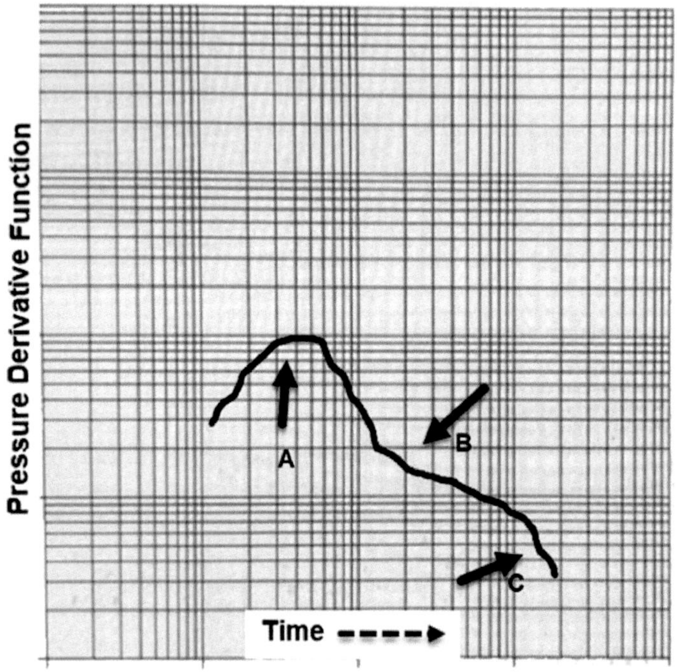

Answer: A: Wellbore storage and skin B: Radial flow C: Boundary effects or approaching a higher transmissivity interval.

Q5-Do a visual inspection of the data shown below and approve or reject the data and explain your reasons.

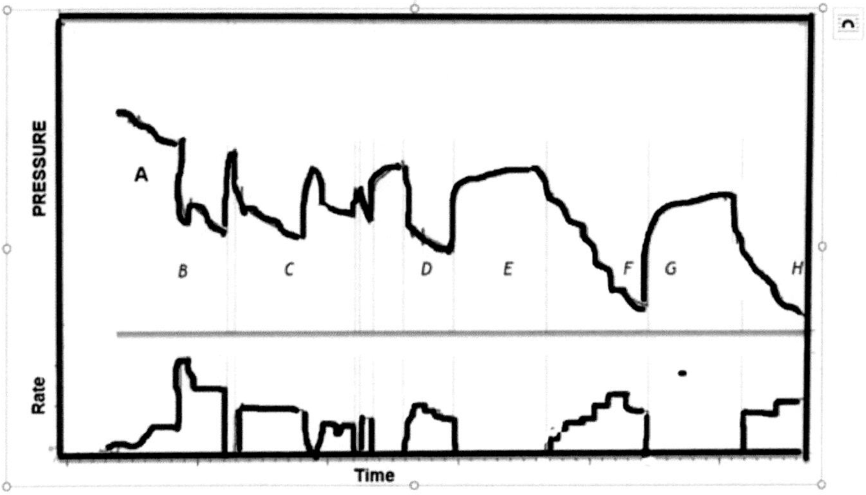

Answer: In intervals, C and F, some of the rate changes do not correspond to the recorded pressure changes.

Q6-Show the expected pressure derivative plots for Wells A and B.

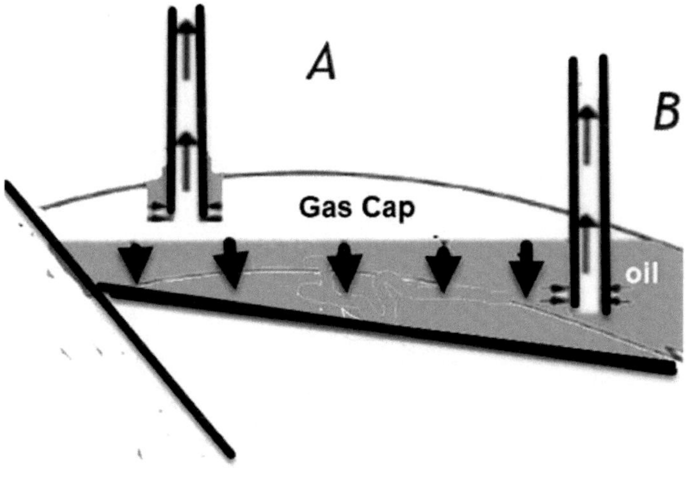

Answer:

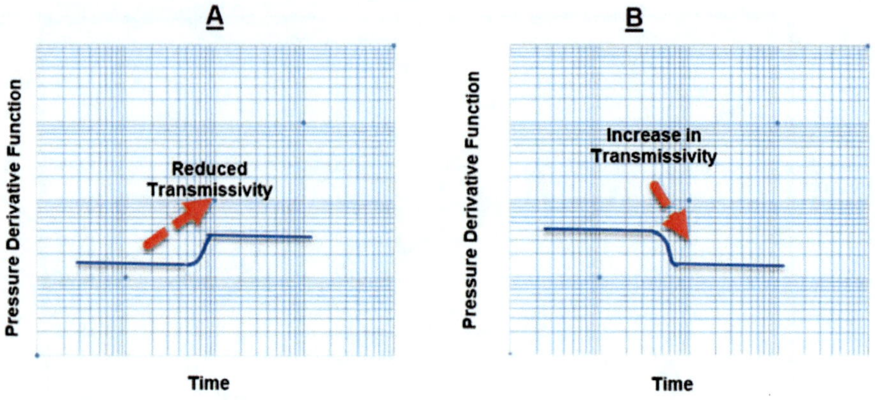

Q7-Show the derivative function plots for the conditions listed below:

a—Wellbore storage, skin, well between two parallel faults.

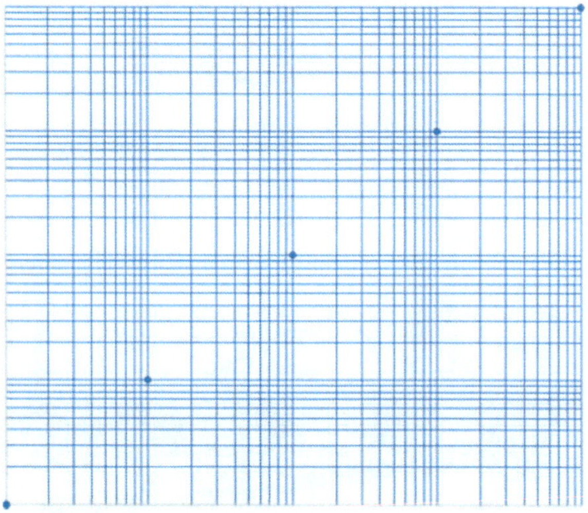

b—Wellbore storage, skin, NFR- unsteady state.

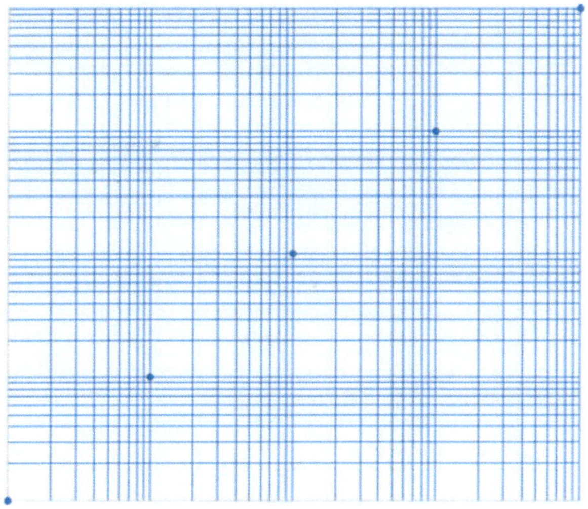

c—Wellbore storage, skin, radial flow, perpendicular faults.

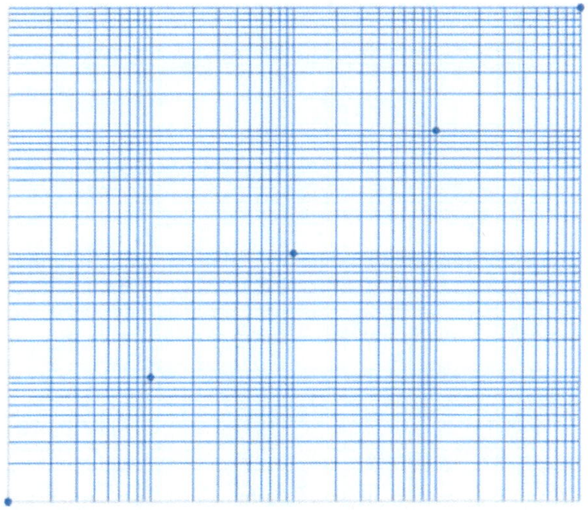

d—No wellbore storage, no skin, radial flow, cross-flow among layers.

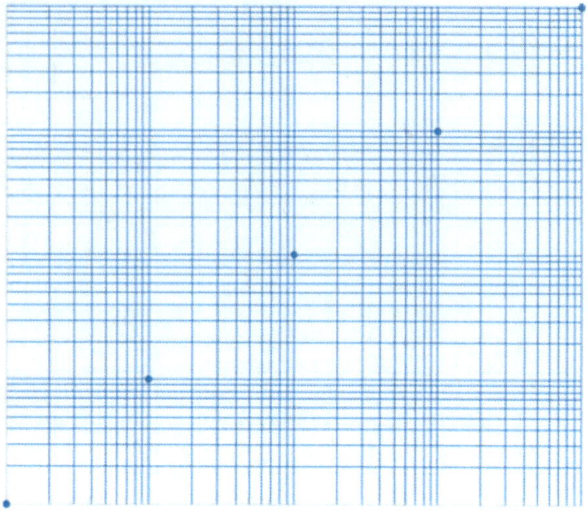

e—No wellbore storage, no skin, choked hydraulic fracture.
f—No wellbore storage, no skin, PSS Naturally Fractured Reservoir.

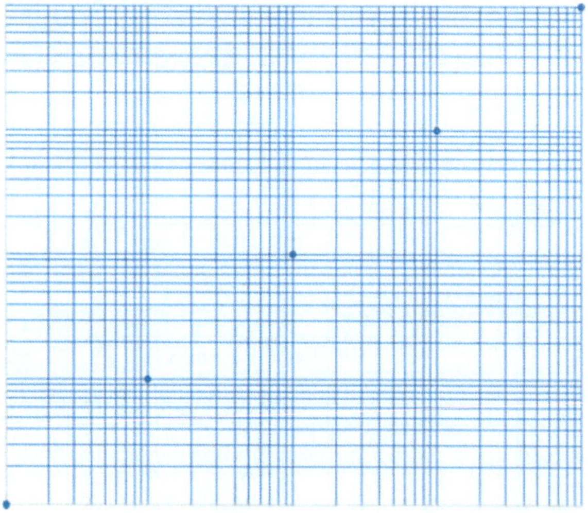

Answers:

a- WS, Skin, well between two parallel faults

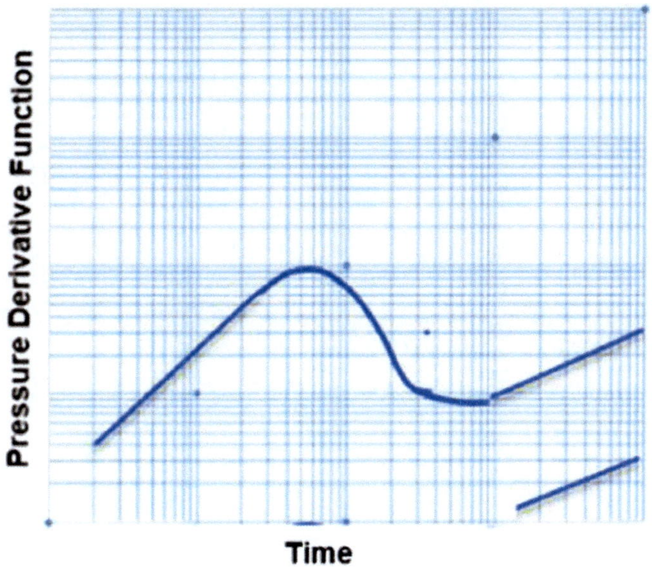

b-WS. Skin. NFR- USS

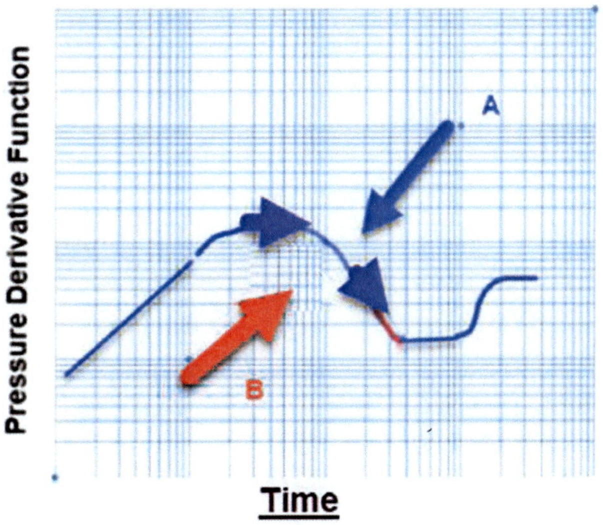

c-WS, Skin. Radial flow, Perpendicular Faults.

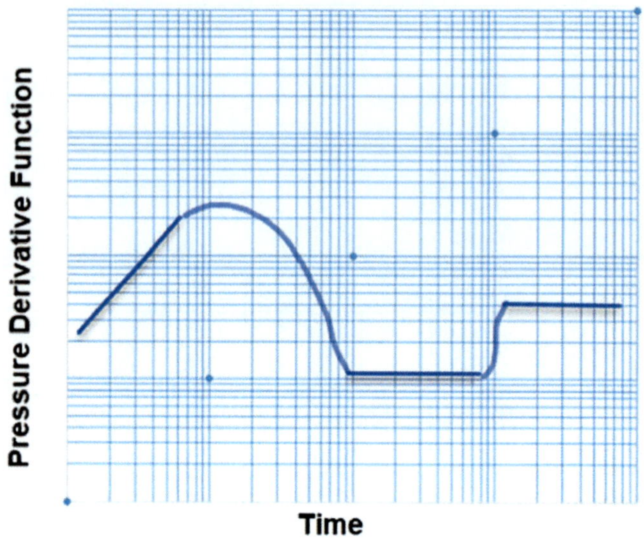

d-No WS. No skin, Radial flow, Cross flow among layers.

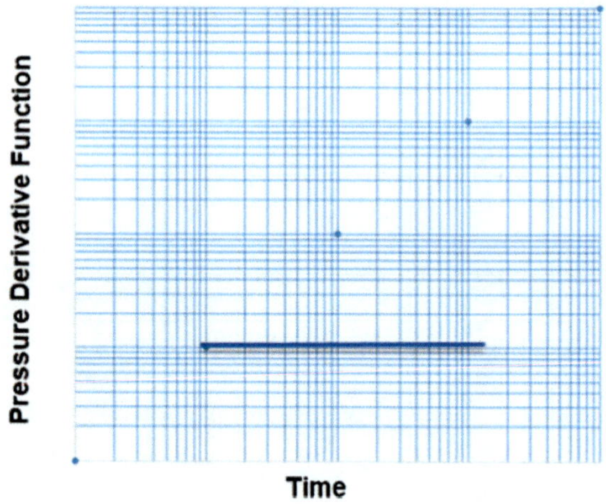

e-No WS, no Skin, Choked hydraulic fracture.

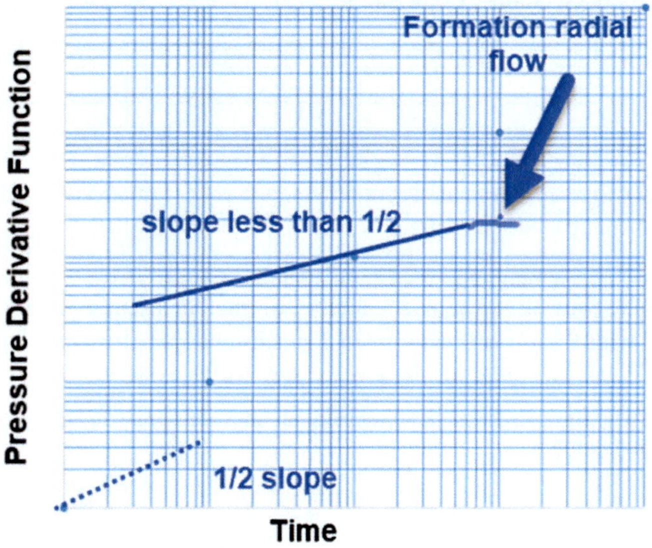

f: No wellbore storage, no skin, PSS Naturally Fractured Reservoir.

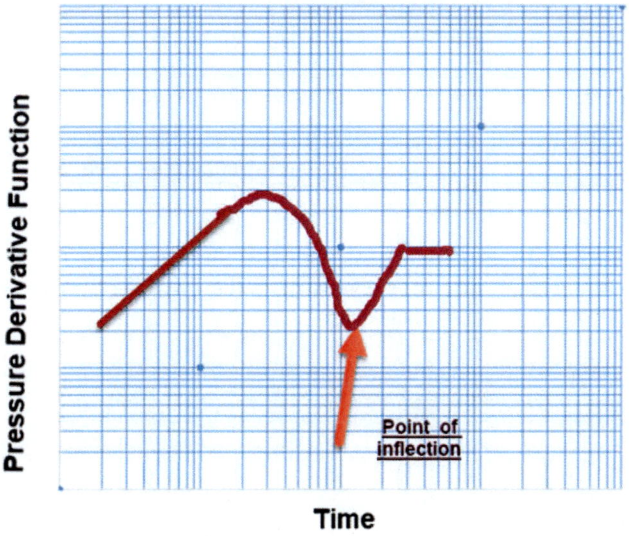

Q8-Back pressure equations using the isochronal testing for 10 gas wells were prepared and they all show the same value of (n) but different values for C. How do you explain that?

Answer: The C term in the back pressure equation is a function of permeability, skin and drainage radius (test duration).

$$C = \frac{\pi kh}{\ln\left(\frac{re}{r_w}\right) - 0.75 + S}$$

Q9-For a steam-dominated geothermal reservoir, if steam is treated as gas, then rather than using pseudo pressure, one may use P or P^2. Explain.

Answer: In testing gas reservoirs, pseudo pressure is used in the representation of pressure for well test analysis. It turns out that the pseudo-pressure is linearly correlated with P^2 at low pressures. In a steam-dominated reservoir, the pressure is low, and P^2 can substitute for pseudo pressure for analysis.

Q10-What kind of test is shown below, and what can you estimate from this test? Show your estimate.

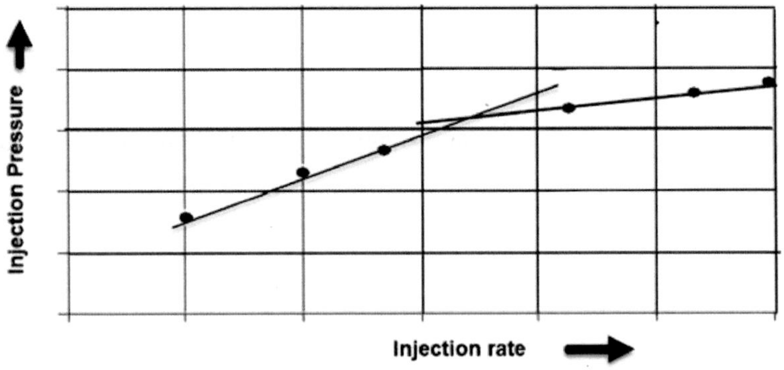

Answer: A step rate test. We may estimate the pressure that causes the fracturing pressure of the formation under test.

Q11-What type of flow is indicated in the section marked by the arrow? What can be the cause of this behavior?

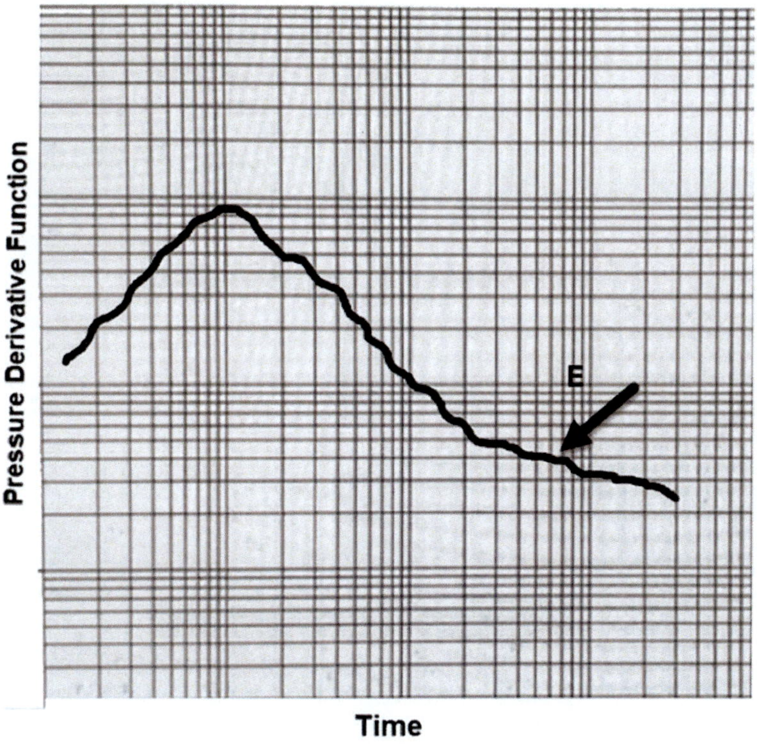

Answer: The Upper-pressure derivative plot starts with evidence of wellbore storage and skin and exhibits a hemi spherical flow with a potential radial flow afterward. For the lower plot, assuming a non-horizontal well, we see wellbore storage and skin followed by radial flow at B, and period A that can be indicative of a fault.

Q12-Consider the following equation where Alpha is a constant. What type of flow is represented? Derive an expression for the derivative function plot.

$$P = \alpha * t^{1/4}$$

Answer 1: Bilinear flow.

Answer 2: The pressure derivative function is:

$$P'\,dt = \frac{\alpha}{4} t^{\frac{1}{4}}$$

Q13-Given the following well and reservoir flow setting, show a pressure derivative function plot representing the expected results.

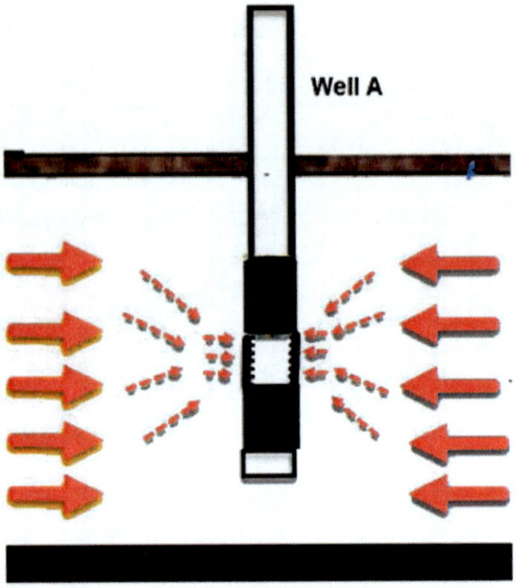

Answer:

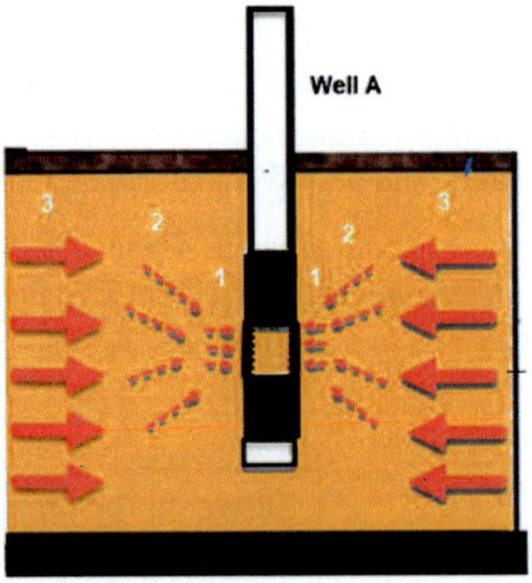

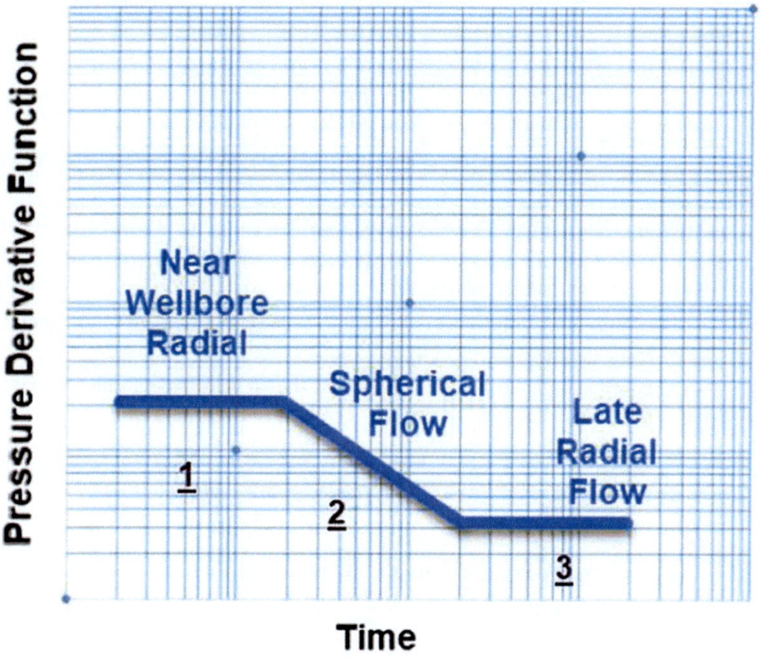

Q14-Two wells are located 200 ft. from each other. Starting from a static condition, one well is placed on production, and the pressure change in the other well is measured. Data recorded follows the Theis Curve. After 20 h, the data recorded shows higher pressure drops than that expected from the Theis curve. Explain the reason.

Answer: The interference of another producing well can influence the recorded reservoir pressure.

Q15-What causes the skin factor measured in a buildup test of gas well to show production rate sensitivity?

Answer: The skin factor measured in the buildup test of gas wells is a function of the flow rate before the buildup test. At higher rates, the pressure drop increases because of turbulence, resulting in a higher value for the skin factor.

Q16-What will be the equivalent radius of a well if the actual radius of the well is 0.5 ft. with negative skin of – 2.

Answer: $r_{eq} = r_w * e^{-s} = 0.5 * e^{+2} = 3.69\text{ft}$

Q17-Indicate True or False: T, F

True or False	Statement
	1—A subterranean formation at a depth of 2000 ft contains Uranium minerals. The best way to way to evaluate the in-place amount of Uranium is pressure buildup.
	2—A combination of pressure drawdown and pressure buildup can help in identifying stress-sensitive formations.
	3—Permeabilities obtained from core plugs are always the same as those obtained from pressure buildup.
	4—In formations subjected to hydraulic fracturing, the flow regime is primarily radial.
	5—For a thin formation where the perforated intervals covers the entire thickness, the flow regime is spherical.
	6—The format of the diffusivity equation as shown below applies to x, y, and z system and is not for a radial flow system is: $\nabla p2 = \frac{\varphi\mu c}{k}\frac{\partial p}{\partial t}$
	7—The following condition implies a steady state condition. $\frac{\partial P_D}{\partial t_D}$ = constant
	8—Under steady-state conditions, the diffusivity equation is elliptic.
	9—In the following format of the flow equation, the pressure drop caused by the skin is (0.87 S). Pi -Pwf(t) $= [\frac{162.6q\mu B}{k.h}[\log(t) + \log[\frac{k}{\varphi\mu crw^2}] - 3.23 + 0.87\ S]$

Answers:

1	**False:** There is no expected fluid flow dynamics in a Uranium bearing formation.
2	**True:** There are expected changes observed in estimated permeability between the two tests.
3	False: Core permeabilities represent a small fraction of the reservoir surrounding a well.
4	False: If a fracture has been developed the flow will be primarily through the fractures and the pressure response represents linear flow.
5	False: The flow is expected to be radial.
6	True:
7	True
8	True: Elliptic equations represent conditions that do not change with time.
9	False: Pressure drop caused by the skin is 0.87*m*S where: m $= \frac{162.6q\mu B}{k.h}$

Q18-In the following test sequence, the questionable measurement is:

A—q1

B—q2

C—q3

D—They all correlate.

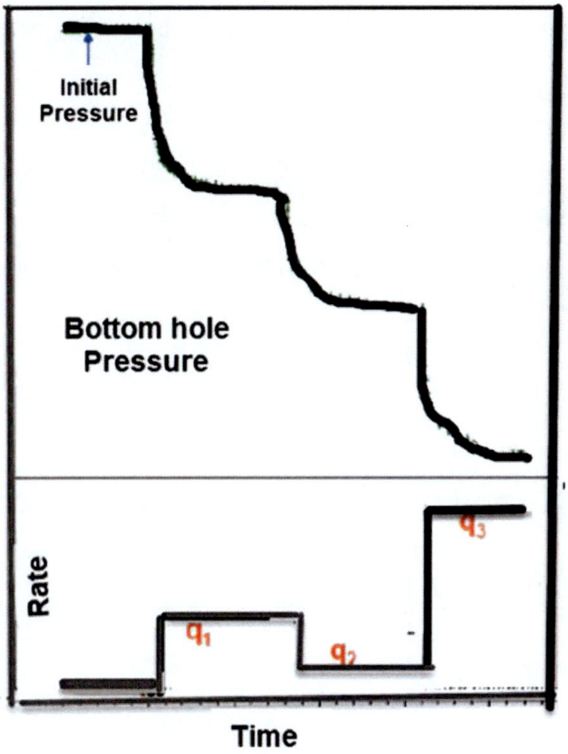

Answer: q2, as the expected pressure rise is not realized. Pressure readings correlate well for q1 and q3.

Q19-The following equation describes the pressure response at an observation point at a distance of (r) from a producing well. Given all the same properties of reservoir rock, the same volumetric flow rate for the producer, for which type of reservoir the pressure drop will be higher:

$$P(\mathrm{r}, \mathrm{t}) = P_i - \frac{70.6 q \mu B}{kh}\left\{-E_i\left[-\frac{948 \varphi \mu c r^2}{kt}\right]\right\}$$

A—Very heavy oil reservoir
B—Very light Oil reservoir
C—No difference

Answer: Heavy oil reservoirs representing oil of high viscosity will show further pressure drop *as*:

$$\frac{70.5\mathrm{q}B\mu}{\mathrm{kh}} \quad \text{Increases with } \mu\text{, the viscosity.}$$

Q20-The expression for the pressure drop caused by skin factor (s) can be written as:

A—$\Delta p_{skin} = 162.6\frac{q_{sf}\mu}{kh} \cdot S$
B—$\Delta p_{skin} = 141.2\frac{q_{sf}\mu}{kh} \cdot S$
C—$\Delta p_{skin} = 0.87\frac{q_{sf}\mu}{kh} \cdot S$

Answer: B is the answer. as $\nabla pskin = 0.87(\frac{162.6\ qB\mu}{kh})S$

Q21-If the sand face flow rate is 2400 STB/D and the wellbore storage is trivial, the surface flow rate is:

A—The same
B—Larger
C—Smaller

Answer: A is the answer.

Q22-The P'. dt derivative function expression for the following equation may be written

$$P(t) = p_i - \frac{162.6q\mu B}{kh}\left\{\log(t) + \log\left[\frac{k}{\varphi\mu c r w^2}\right] - 3.228 + 0.8686.\ \text{Skin}\right\}$$

as:

A—$\frac{-162.6qB\mu}{kh}$.
B—$\frac{-141.2qB\mu}{kh}$.
C—$\frac{-70.6qB\mu}{kh}$.
D—None of the above

Answer: $P' * dt = -\frac{162.6qB\mu}{kh} * \frac{t}{t*Ln(10)} = \frac{-70.6qB\mu}{kh}$.

Q23-See the following calculations. Without doing any computations, and assuming all other parameters being the same, what would be the impact on estimated bottom hole pressures if the gas gravity were 0.65?.

Parameter:
Gas Specific Gravity 0.55
Tubing internal diameter 3 in Relative roughness 0.0002
Flow rate 1 MMSCF/D
Pipe length 5000 ft
Hole angles 2° Tubing head pressure 2000 psia
Bottom hole temperature 150 F
Bottom Hole Pressure 2223 Psia

Answer: Higher gas gravity affects the column pressure and the estimated bottomhole pressure will be higher.

Q24-The Horner plot for a well with negative skin will look like this:

A

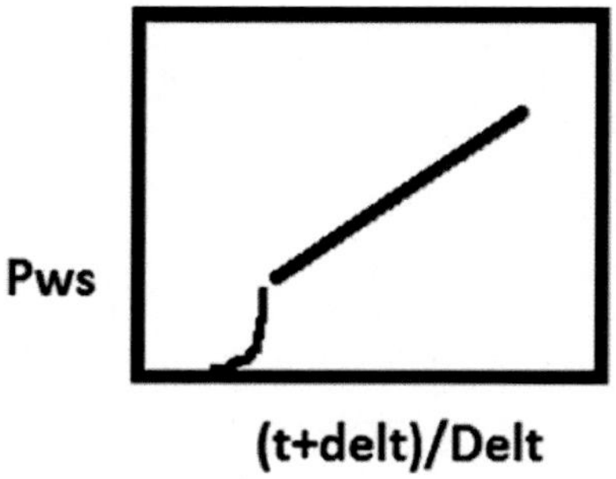

B

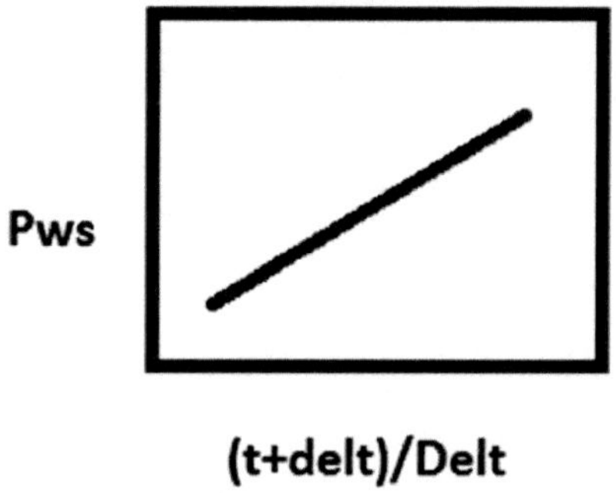

C

Answer: C

Q25-Given the following and using the type curve of Agarwal et al (1979) estimate fracture conductivity.

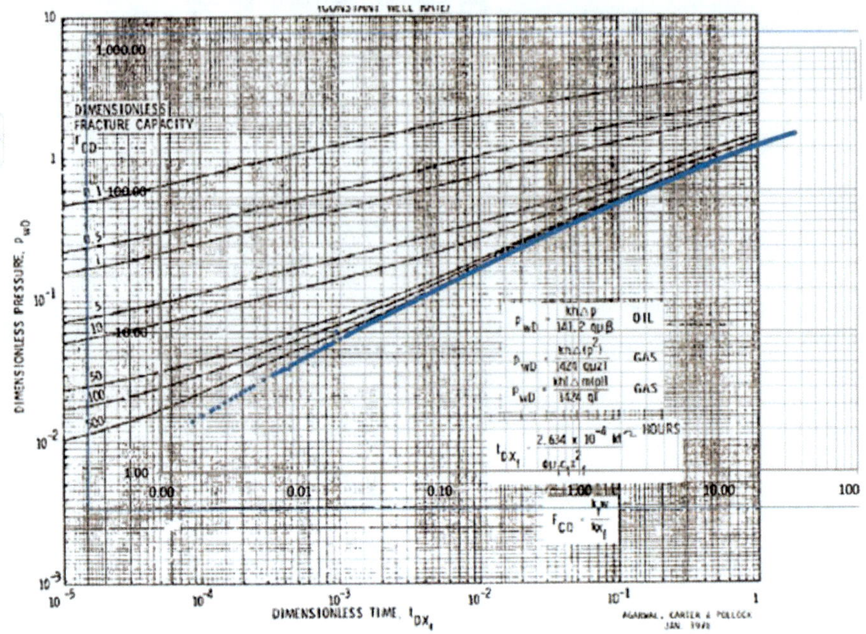

(Courtesy SPE 6838)

Answer: We can match the pressure vs. time data on a log-log scale with a fracture conductivity of 500.

Q26-An engineer who was analyzing a buildup test for a horizontal well quits his job and left the following graphs. Your supervisor wants you to suggest a method to complete his job and use these graphs and estimate Kx, Ky and KZ.

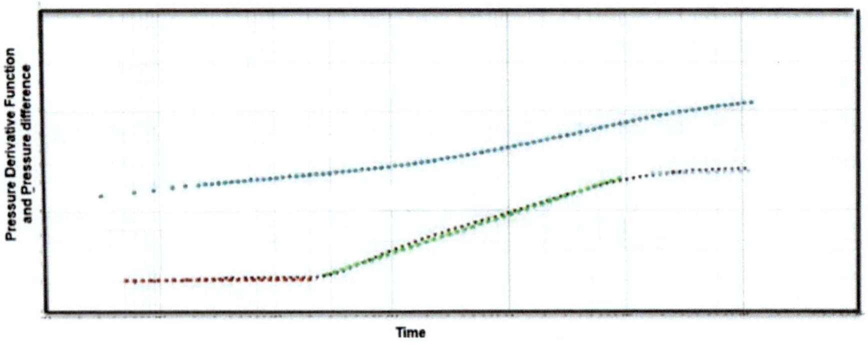

Answer: We can estimate Kx * Kz from a semi-log plot of P versus early time data. From the derivative function plot, the intersection with the linear flow period can help to estimate Kz. Ky can be estimated from a cartesian plot of pressure vs. $\sqrt{t}$. For the linear flow period. If the late-time pseudo radial flow develops, the semi-log plot vs. time can give another estimate of Ky.

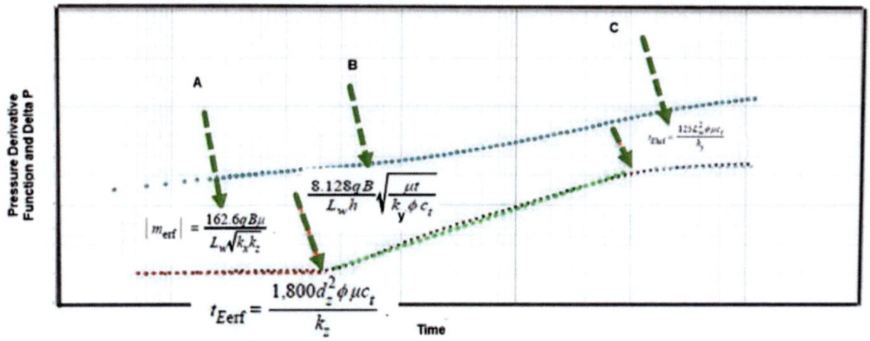

This is based on assuming that we know the true lateral length of the horizontal well.

If lateral length is also not known, the solution of four unknown, Kx, Ky, Kz, and L, requires using an iterative nonlinear estimation method. Formulations are shown below:

From the early radial flow

$$Pi - Pwf = \frac{162.6q\mu B}{Lw\sqrt{kz}\cdot kx}\left\{\log\left(\frac{\sqrt{kxkzt}}{\varphi\mu ctrw^2}\right) - 3.227 + 0.868\ \mathrm{S}\right\}$$

$$\text{Slope of semi log plot early radial flow} = \frac{162.6q\mu B}{Lw\sqrt{kz}\cdot kx}$$

$$\text{Time to the end of early radial flow} = \frac{1800d_z^2\varphi\mu ct}{Kz}$$

The linear period:

$$Pi - Pwf = \frac{8.128qB}{Lw\cdot h}\left(\frac{\mu.t}{\sqrt{kx}\cdot\varphi\cdot ct}\right\} + \frac{141.2q\mu B}{Lw\sqrt{kx, kz}}(\mathrm{S})$$

Q27-If the straight line shown is the correct line representing the middle time region on a Horner plot, and the producing time was 200 h, estimate the wellbore storage coefficient Pwf = 3380 psi q = 200 STB/D B = 1.2

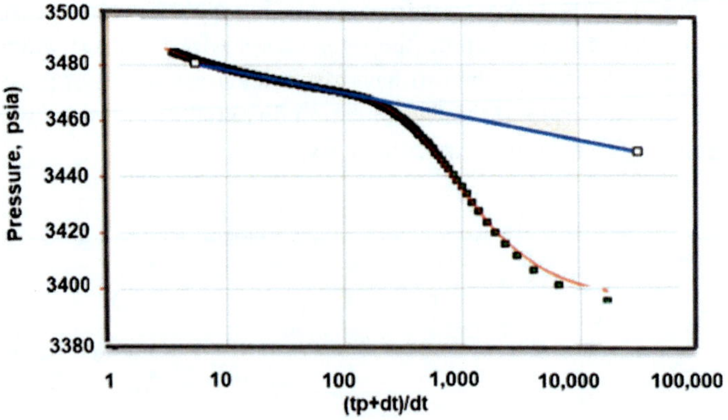

Answer: We can use one of the early points and estimate the corresponding dp and dt.

If (t + dt)/dt = 1000, for t = 200 h. it corresponds to dt = 0.2 hrs.
dp = 3435 − 3380 = 55 psi. Then the wellbore storage coefficient is

$$\mathrm{CW} = \frac{\mathrm{q} \cdot \mathrm{B}}{24} * \left(\frac{dp}{dt}\right) = \frac{(200)(1.2)}{24} * \left(\frac{0.2}{55}\right) = 0.036\ bbls/psi$$

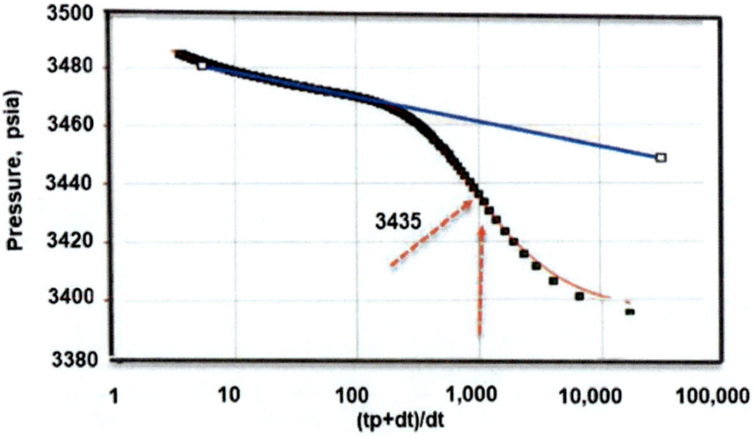

Q28-Given the following data, write the equations and estimate the wellbore pressure after 24 h of flow if the reservoir is infinite acting.

Initial Shut-in pressure = 3000 psi. Assume no skin. Initial rate for the first 12 h is 500 STB/D, rate for the next 12 h is 1000 STB/D, reservoir transmissibility = 4300 mD. ft./hr., Reservoir storativity = 0.0015 ft./psi, rw = 0.5, B = 1.2.

$$\text{Pwf} = \text{Pi} + 70.6\frac{q_1.\mu.B}{k.h}\left\{\ln\frac{1688\varphi.\mu.ct.r_{w^2}}{k.t}\right\} + 70.6\frac{q_2.\mu.B}{k.h}\text{Ei}\left\{\frac{948}{k.t}\varphi.\mu.ct.r_{w^2}\right\}$$

$$\text{Pwf} = \text{Pi} + 70.6\frac{q_1.B}{T}\left\{\ln\frac{1688.S.r_{w^2}}{T.t}\right\} + 70.6\frac{q_2.B}{T}\text{Ei}\left\{\frac{948}{T.t}.S.r_{w^2}\right\}$$

Answer:

$$\text{T} = \frac{\text{k.h}}{\mu} = 4300\ \text{mD.ft./h}$$

$$S = \phi.\text{ct.h} = 0.0015\ \text{ft./psi}$$

$$\text{t} = 24\ \text{hrs.}\ \text{t}_1 = 12\ \text{hrs.}$$

$$Pwf = \text{Pi} - \frac{70.6 * q1 * B}{\text{T}}\left(\ln\left(1688\left(\frac{S}{T * t}\right)rw^2\right)\right) - \frac{70.6 * (q2 - q1) * B}{\text{T}}\left(\ln\left(1688\left(\frac{S}{T * (t - t1)}\right)rw^2\right)\right) = 2759.65\ \text{psi}$$

Q29-Given the following data, estimate the pseudo skin caused by partial completion. Total formation thickness = 200 ft., length of completed interval from the top = 100 ft. rw = 0.5 ft., kv = 10 mD, kh = 100 mD.

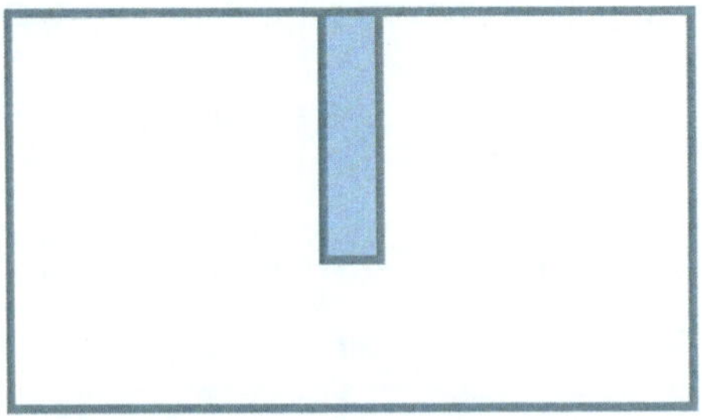

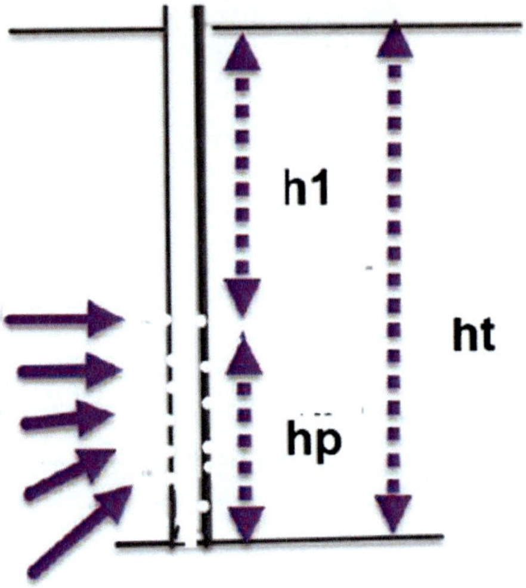

Using $h_{1D} = h_1/h$ $h_{pD} = h_p/h$

$$A = \frac{1}{\mathrm{h1D} + hpD/4}$$

$$B = \frac{1}{h1D + 3hpD/4}$$

$$r_D = \frac{rw}{h}\left(\frac{kv}{kh}\right)^{0.5}$$

$$Sp = \left(\frac{1}{hpD} - 1\right) Ln\frac{\pi}{2rD} + \frac{1}{hpD} Ln\left[\frac{hpD}{2 + hpD}\left(\frac{A-1}{B-1}\right)^{0.5}\right]$$

*Papatzacos, P.:1987 Approximate partial penetration pseudo skin for infinite conductivity wells SPE 13956 SPE Reservoir Engr.227–234

$$\text{Sp} = \left(\frac{1}{hpd} - 1\right)\ln\frac{\Pi}{2rd} + \frac{1}{hpd}\ln\{\frac{hpd}{2+hpd})\left(\frac{A-1}{B-1}\right)^{0.5}\}$$

Where $h_{1d} = h1/h$ $h_{pd} = hp/h$ $A = \dfrac{1}{h1D + hpd/4}$ $rd = \dfrac{rw}{h}\left(\dfrac{ky}{kh}\right)^{0.5}$

$$B = \frac{1}{h1D + 3hpd/4}$$

$h_{1D} = 0/200 = 0.$ hpd $= 100/200 = 0.5$ $A = \dfrac{1}{0 + 0.5/4} = 8$

$$B = \frac{1}{0 + 1.5/4} = 0.266$$

$$rd = \frac{0.5}{200}(.1)^{.5} = 0.000$$

$$Sp = (2-1)\text{Ln}\left(\frac{\pi}{2 * 0.00079}\right) + \left(\frac{1}{0.5}\right)\ln\left(\frac{0.5 * \left(\frac{7}{-0.734}\right)^2}{2+0.5}\right) = 13.4$$

Q30-From a static condition, three wells A, B and C produce for 24 h and pressure drop is measured in Well D after 24 h. Given the same reservoir conditions, what will be the pressure drop in well D after 24 h if the producing rates for the three wells were doubled from the beginning?

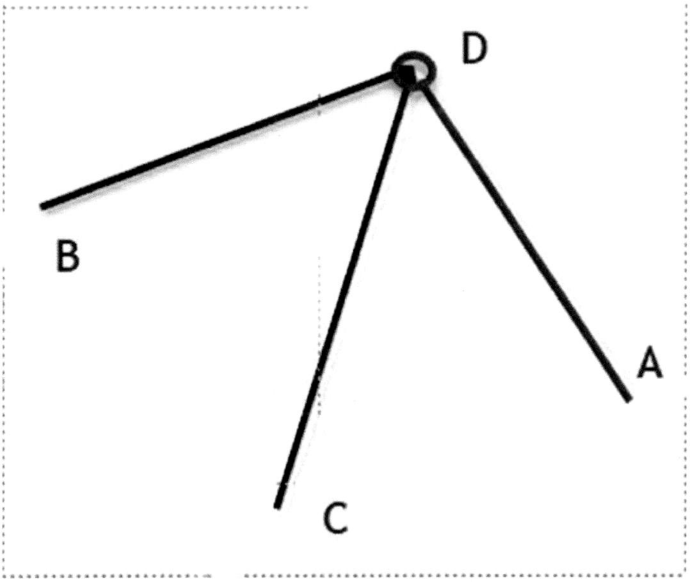

Answer: Pressure droop will be twice as the doubling of the rate (q) doubles the pressure drop according to the following equation:

$$dp = -70.6\frac{q * \mu * B}{k * h} Ei\left(-\frac{948.\phi.\mu.Ct.d^2}{k * t}\right)$$

Q31-The pressure drop caused by skin is 800 psi. If the Horner plot slope m = 50 psi/cycle, estimate the skin factor.

Answer: dp (skin) = 0.87 * m * S thus, s = 18.4

Q32-A well in an infinite acting reservoir is produced at a constant rate of 500 STB /D and the bottomhole flowing pressure after 12 h is 4200 psi. If this same well had produced at 750 STB/D under the same conditions, what would have been the bottomhole flowing pressure? Initial static pressure is 5000 psi.

Answer:

$$Pwf = \text{Pi} - \frac{70.6 * q1 * B}{\text{T}}\left(\ln\left(1688\left(\frac{S}{T * t}\right)rw^2\right)\right)$$

$$4200 = 5000 - \frac{70.6 * q1 * B}{\text{T}}\left(\ln\left(1688\left(\frac{S}{T * t}\right)rw^2\right)\right)$$

$$800 = \frac{70.6 * 500 * B}{\text{T}}\left(\ln\left(1688\left(\frac{S}{T * t}\right)rw^2\right)\right)$$

$$X = \frac{70.6 * 750 * B}{\text{T}}\left(\ln\left(1688\left(\frac{S}{T * t}\right)rw^2\right)\right)$$

$$\frac{800}{\text{x}} = \frac{500}{750} \quad \text{x} = 1200 \text{ psi}$$

Q33-A pressure build up test was conducted for a well producing a single-phase compressible fluid with Ct = 40e^{-6} (1/psi) and at a wellbore storage of 0.1 Bbl./psi. The wellbore diagram is shown below. Compare the wellbore storage with that obtained from the buildup and discuss your findings. Compressibility of the fluid at the mean wellbore pressure and temperature, psi^{-1}in the well system is 50e^{-6} 1/psi.

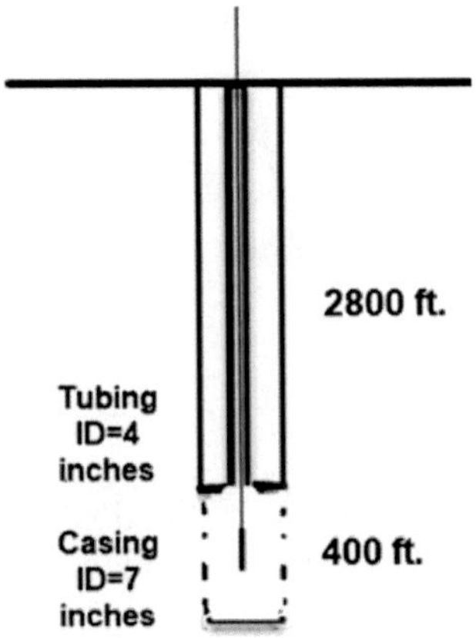

Answer: Tubing capacity of ID = 4 in which is 0.0155 Bbl./ft. For a 2800 ft. interval, it translates to 43.5 Bbl. Casing Capacity for 7 In ID 0.0477.

Bbl./ft. and for a 400 ft. interval, it is 19.09 Bbl./ft. As such, the estimated wellbore storage is:

$$Cs = Vws{*}Cws$$

$C_s = 50e^{-6}*$ (10.69 + 19.09) = 0.00148 Bbl./psi. Because the estimated value from a build-up is higher, indicating other cavities and volume spaces, such as tubing leaks and casing defects, are seen from the buildup test.

Q34-Well A is produced at 500 STB/D for 24 h, and then the well is shut in for 24 h. Well, B is an observation well. Assuming an infinite-size reservoir, what is the expected pressure reading in well B after 24 h? K = 300 md h = 150 ft., ϕ =0.2, r_w = 0.5 ft., Ct = $30e^{-6}$ Viscosity = 1.5 cp, B = 1.2.

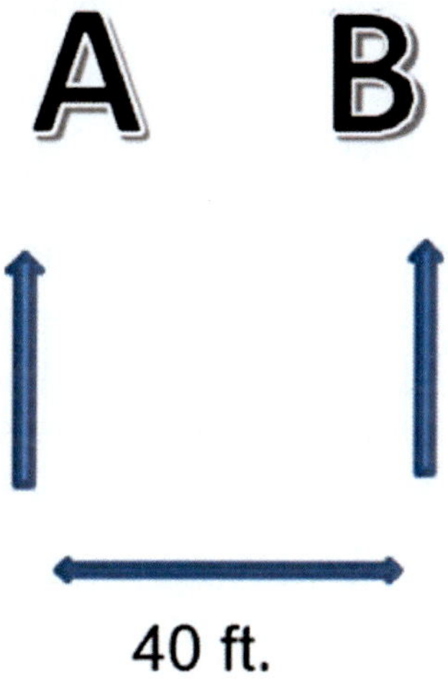

Answer:

$$T = \frac{\mathrm{k} \cdot \mathrm{h}}{\mu} = \frac{300 * 150}{1.5} = 30{,}000.\ mD.ft./cp$$

$$= \Phi * \mathrm{Ct} * \ \mathrm{h} = 0.2 * 30 * e^{-6} * 150 = 0.0009\ \mathrm{ft./psi}$$

dp (after 24 hours)

$$= -70.6\frac{\mathrm{q} * \mathrm{B}}{\mathrm{T}} Ei\left(-\frac{948.\ \mathrm{S} * d^2}{\mathrm{T} * \mathrm{t}}\right)$$

$$= -70.6\frac{500 * 1.2}{30{,}000}Ei\left(-\frac{948 * 0.0009 * 400^2}{30{,}000 * 24}\right) = -1.412$$

$$Ei(-0.1896) = 1.8\text{psi}$$

Q35-For a well in a closed reservoir containing water, with a total daily production of 5000 B/D, it takes 300 days for reservoir pressure to become ½ of initial pressure? How long will it take if the rate was 1250 B/D?

$$q.t = Vp * Ct * (P - Pwf)$$

Answer: Here $q.t$ is constant. Reducing the rate by 4 times increases the time by 4 times.

$$t = 1200 \text{ days.}$$

Q36–A new well produces at a rate of 500 STB/D for 24 h. Estimate the distance investigated at the rate of production. What would be the distance if the rate is doubled?

$$\varphi = 0.2\text{k} = 300\,\text{md} \quad \text{h} = 100\,\text{ft.,} \quad \text{Ct} = 10\,\text{e}^{-6}, \mu = 2\,\text{cp}$$

Answer: $d = \sqrt{\frac{0.00105 * k * t}{\Phi \text{mCt}}} = \sqrt{\frac{0.00105 * 300 * 24}{0.2 * 2 * 10e-6}} = 1374$ ft.
Doubling rate causes no change except it establishes a larger dp at this location.

Q37-In a radial system, how far would a new well producing at a rate of 1500 STB oil per day for 24 h investigate? What would be the impact if the total fluid had a total compressibility of 10e^{-6}?

$$\varphi = 0.25, \text{k} = 500\text{mD., h} = 120\text{ft., r}_\text{w} = 0.5\text{ft., Ct} = 10\text{e}^{-6}, \mu_0 = 5\text{cp}$$

$$d = \sqrt{\frac{0.00105 * k * t}{\phi.\mu.Ct}} = \sqrt{\frac{0.00105 * 500 * 24}{0.25 * 5 * 10e - 6}}$$

d = 1003 ft., if Ct is increased 30 times, $d = 183$ ft.

Q38-Consider a well located at a distance of L from a sealing fault. Write an expression for boundary condition at point A.

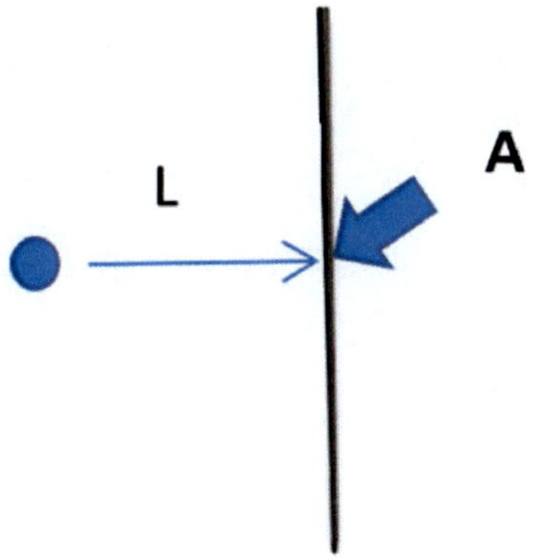

Answer: No flow or (dp/dx) = 0.

Q39-Draw the expected shapes of the Horner plot and the derivative function plot for the conditions shown below. Specify the axis.

a—Vertical hole, wellbore with storage, skin, sufficient test duration infinite acting.

b—Vertical hole, no wellbore storage, no skin, sealing fault cutting through the wellbore.

c—Vertical hole no wellbore storage, no skin, hydraulically fractured well and choked fracture.

Answer: Solutions for conditions a, b and c are shown below:

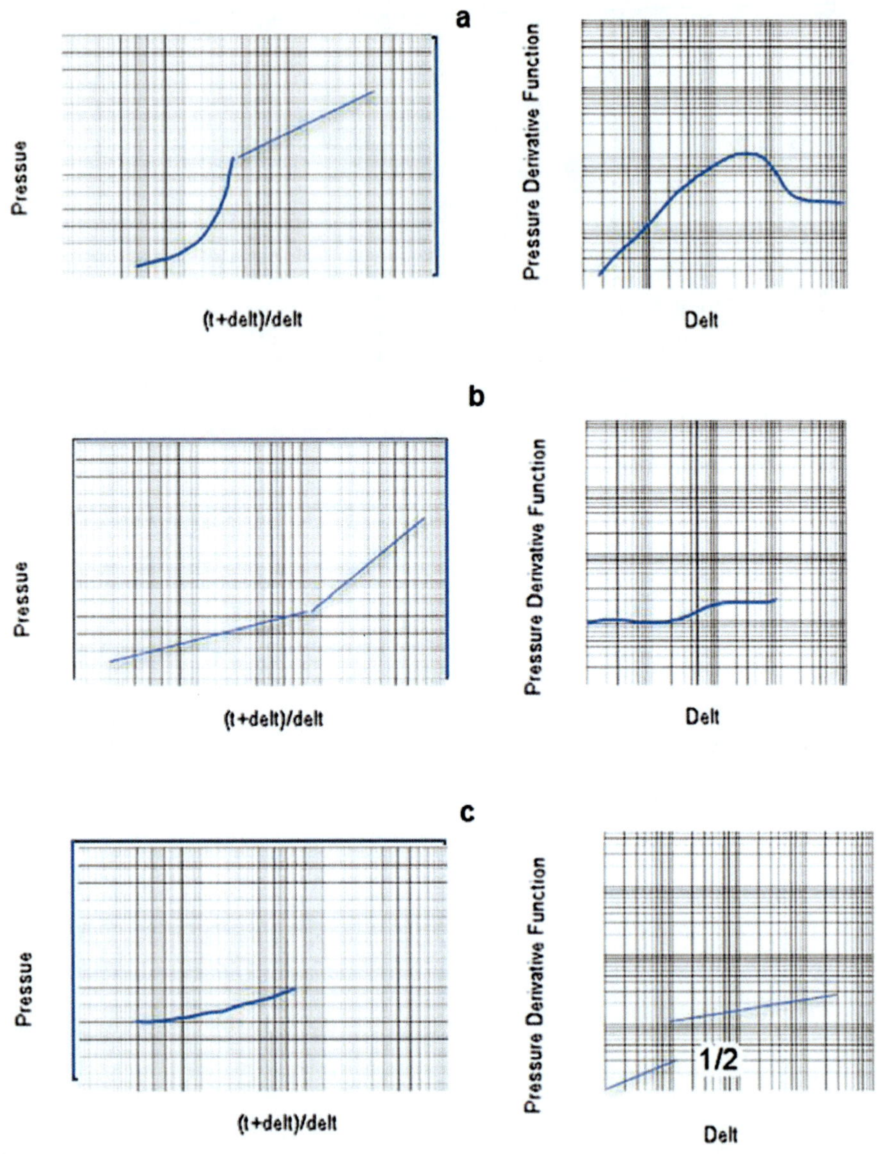

Q40-Identify the conditions when you can observe the following shapes on the derivative plot.

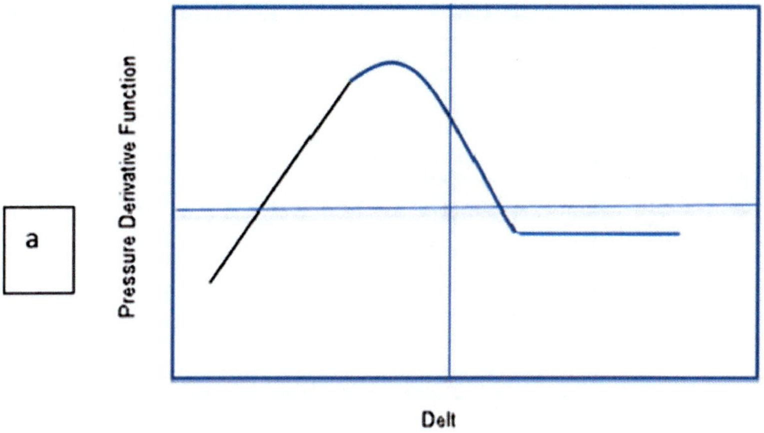

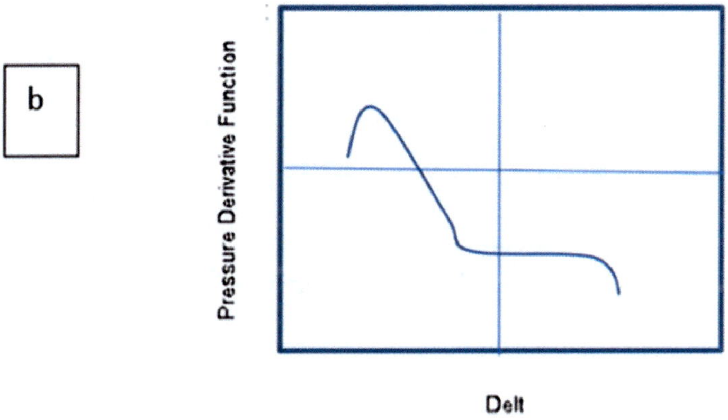

Answer: (a) corresponds to an infinite acting radial flow conditions with wellbore storage and skin, (b) corresponds most likely to bounded radial flow conditions with wellbore storage and skin.

Q41-Identify the conditions when you can observe the following shapes on the pressure derivative function plot for a naturally fractured reservoir:

(a)

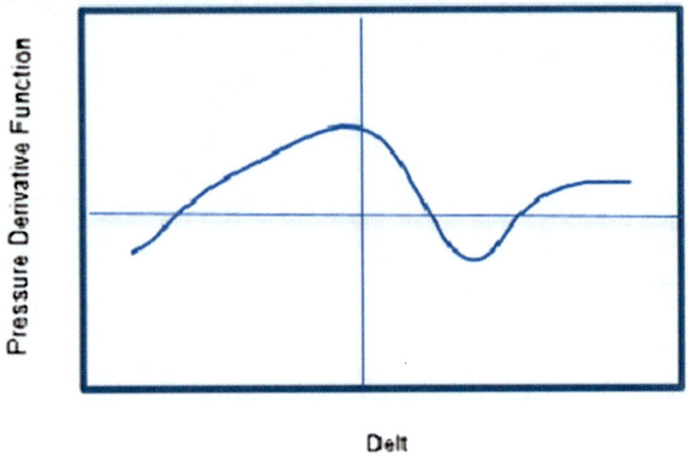

(b)

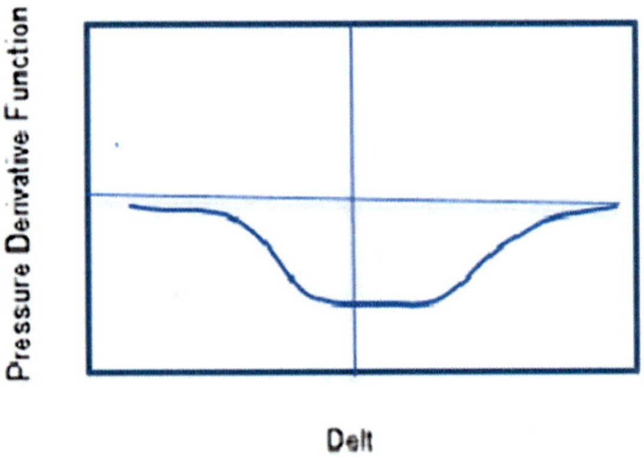

Answer: (a) This corresponds to a well affected by after flow that has masked the initial development of a fracture dominated radial flow. (b) corresponds to a well free of wellbore storage effect by the interporosity effect which is of the gradient type.

Q42-For a closed reservoir containing only oil, total daily oil production is 10,000 STB/D. If the hydrocarbon pore volume is one hundred million barrels, the total compressibility is 50e^{-6} 1/psi, and Bo = 1.2 RB/STB, what is the expected pressure drop after one 5 days production assuming semi steady state?

Answer: $q.t = Vp * Ct * (dP)$

$$dp = \frac{10{,}000 * 1.2 * 5}{10^8 * 50 * 10^{-6}} = 12\ psi$$

Q43-Write a procedure for estimating T and S from the following pulse test data.
$r_w = 5$ ft., B = 1.2

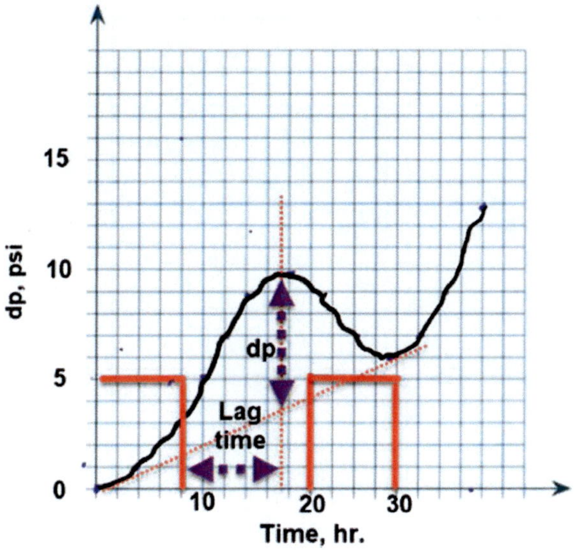

Answer: dp = 6 psi, Time lag = 9.5 h, Cycle time = 20 h, Pulse time = 8 h.

Dimension time lag = (9.5)/20 = 0.475 Ratio of pulse time to cycle time = 0.4

From the type of curve of dimensionless pressure we estimate T and from the dimensionless time chart we estimate S.

Q44-The Horner plot for the DST test on discovery well is shown below. After a shut in of 15 days with an overbalanced mud, a DST was conducted. The well flowed at only 8MMScf/D and the field was undeveloped for several years because it was not thought to be commercial. What conditions could have affected the situation?

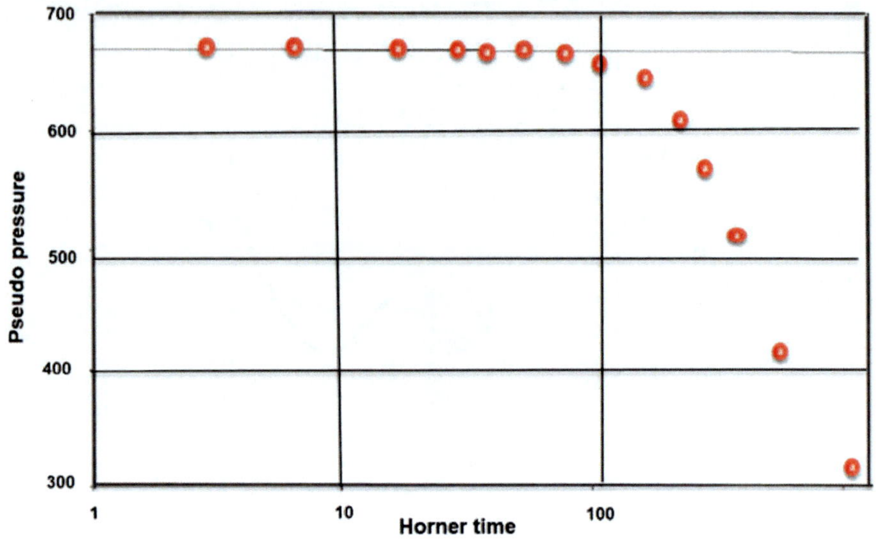

Answer: Damage caused by the overbalanced mud may have contributed to low flow rates.

Q45-The buildup portion of a DST for an offshore well shows the following. Explain the potential causes.

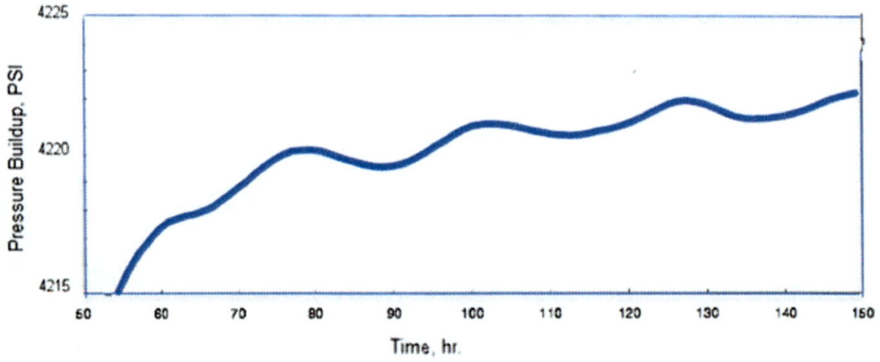

Answer: The effect superimposed on the pressure data is the effect diurnal tides.

Q46-An interference test was conducted between two wells which were 400 ft. apart. The active well produced at a rate of 350 STB/D (B0 = 1.2), rw = 0.5 ft. Write a procedure for estimation of T and S.

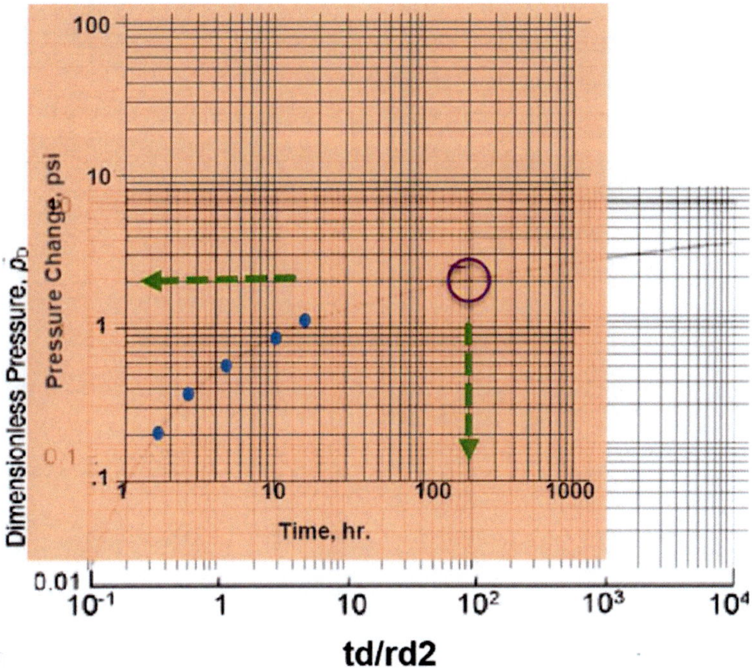

Answer: From a match point (can be any points on the plot of data superimposed on a Theis curve) we can obtain td/rd^2 (dimensionless time and Pd (dimensionless pressure). From Pd and dp retios we estimate T and from td and dt relations we estime S.

Q47: For water production at a rate of 0.5 m^3/s from a groundwater well, the interference at a nearby well located 50 m away is measured and is shown on a Theis curve. Estimate T and S.

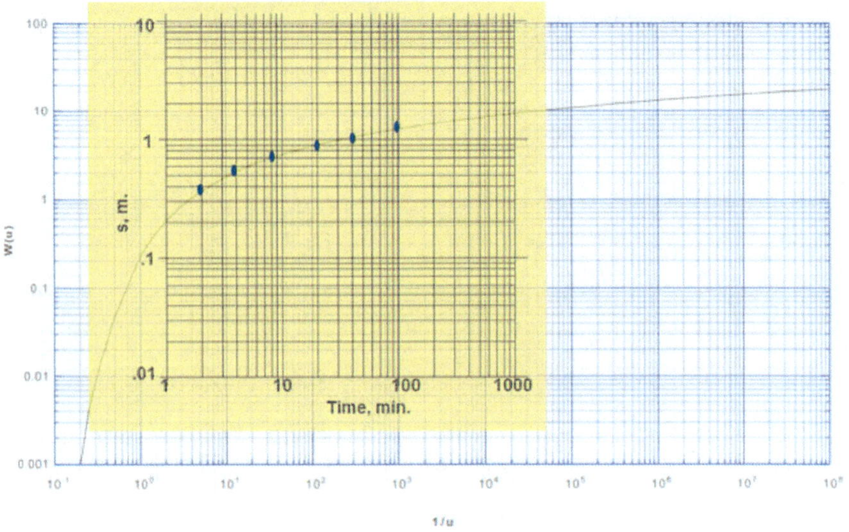

Answer: Match Point

1/u = 1000, W(u) = 6, t = 100 s, s = 1

$$S = \frac{Q}{4\pi T}$$

$$T = \frac{Q \cdot w(u)}{4t * S * \pi}$$

$$T = \frac{5 * 6}{4 * 100 * 1 * 3.14} = 0.0238\text{m}^2/\text{s}$$

$$0.0238 = \frac{5}{4 * 3.14 * S} \quad S = 1.672$$

Q48-A 1500 ft horizontal well was tested while producing at 1000 STB/D and then it was shut in and the following shows the pressure derivative plots for the test. What affects the duration of the segment between the two arrows?

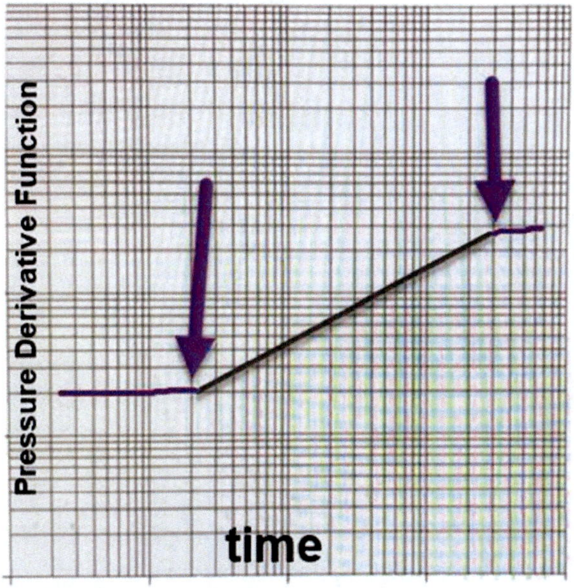

Answer: The length of the lateral.

Q49-What can you conclude from the series of DSTs conducted on a naturally fractured reservoir with the following observations.

DST's only showed flow when a mud acid job was done prior to the DST.

Answer: Fractured were plugged by the drilling mud and the mud acid cleared the blockage.

Q50-Show the log-log plot of dp versus dt for a pressure buildup with a continuously declining wellbore storage.

Answer:

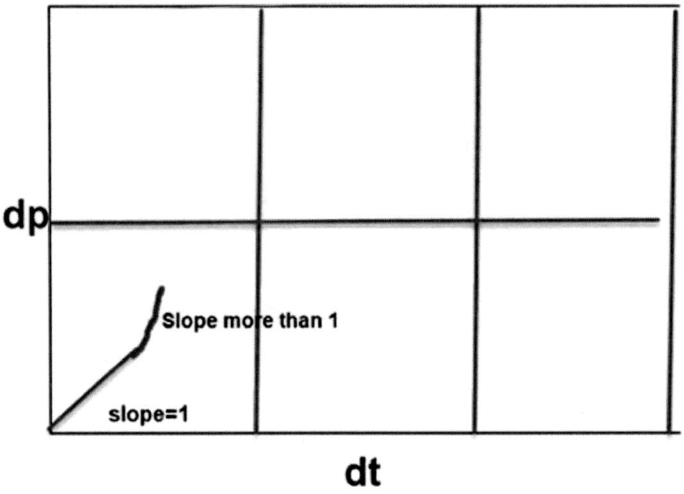

Q51-A buildup test was conducted on a two-layer system and the following data were obtained. What can you say about the reservoir and why?

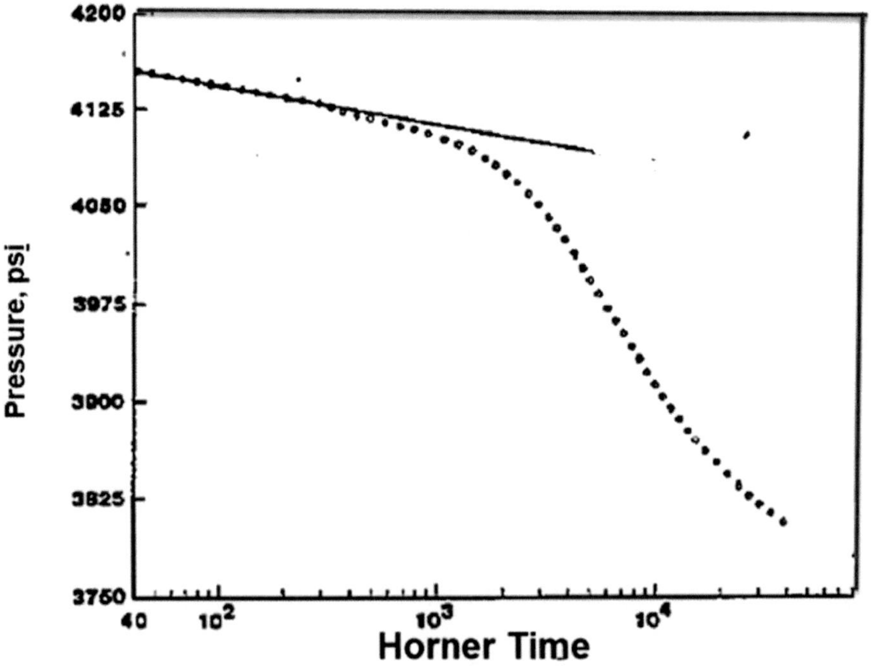

Answer: The test indicates the behavior of one layer and in all likelihood there is strong cross flow between the two layers.

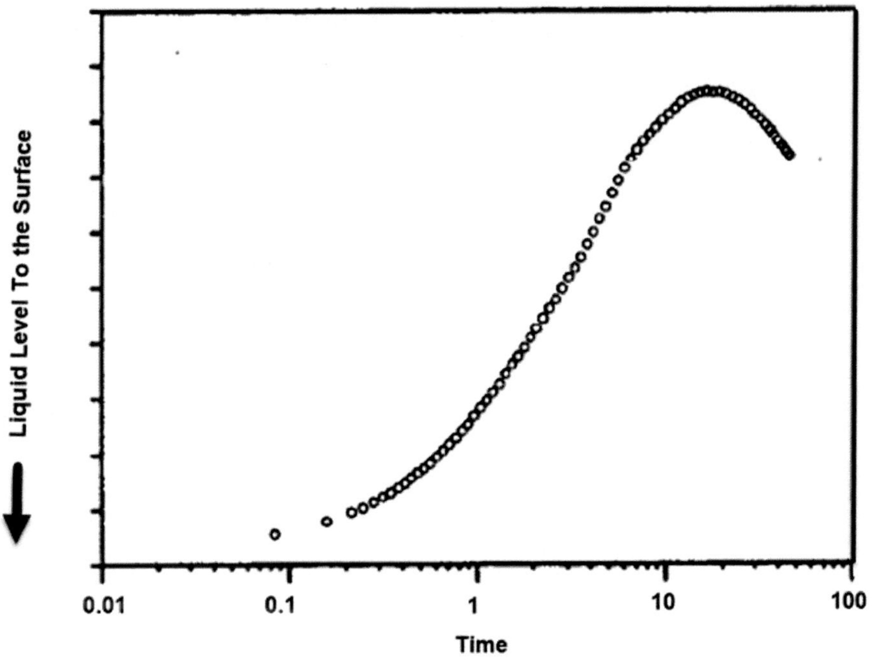

Q52-What is the cause of the above observation in a pumping well test.

Answer: Phase separation in the tubing.

Q53-Estimate fracture conductivity and indicate a procedure to estimate Xf

Answer: FCD is from the matched graph = 10. From a match point and relations between Pd and dp and t_D and dt we can obtain Xf.

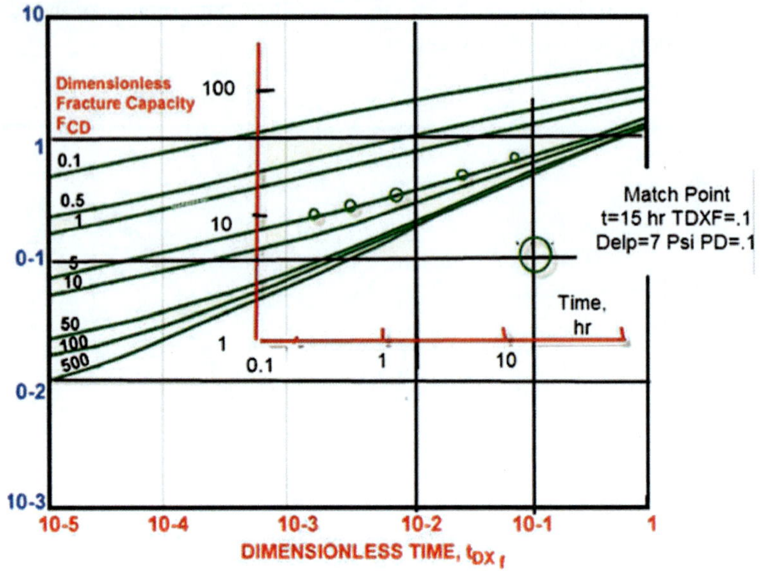

Courtesy Agarwal, Carter and Pollack

Q54-Why could pressure at A be higher than Point B?

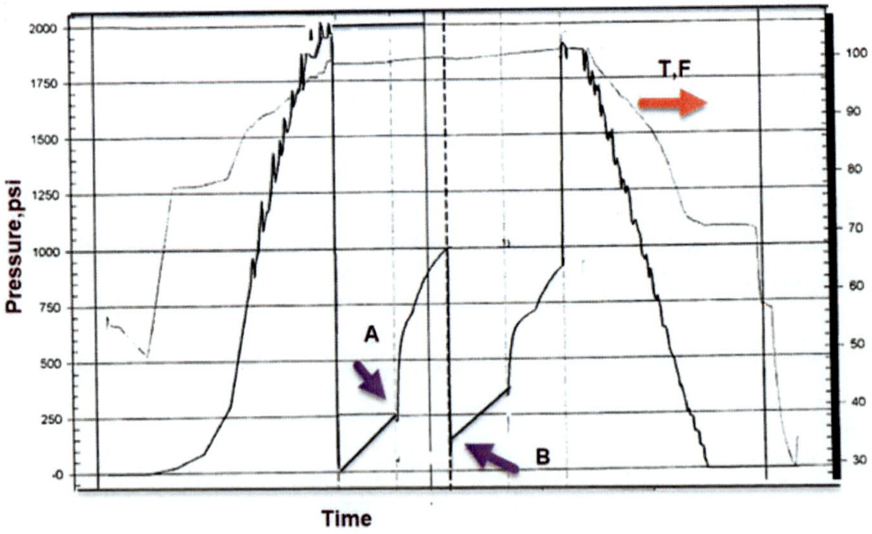

Answer: Reduction of hydrostatic pressure in the drill pipe.

Q55-Explain the shape of the following derivative plot.

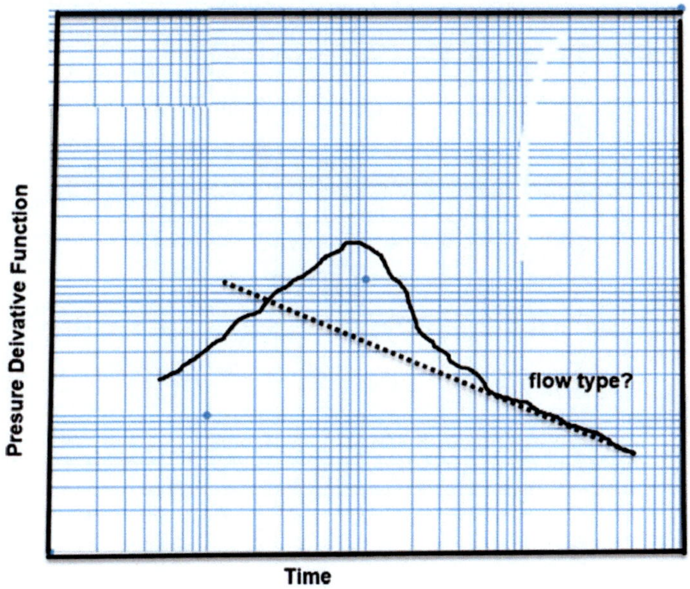

Answer: With a slope of − 1/2, it may be a result of spherical flow caused by partial completion.

Q56-There is evidence that in a producing gas well, condensate has formed, Which log-log plot of pressure derivative function versus time represents the condition.

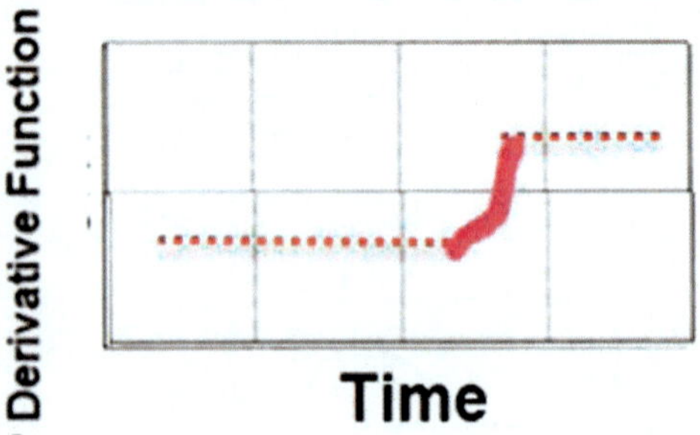

A:

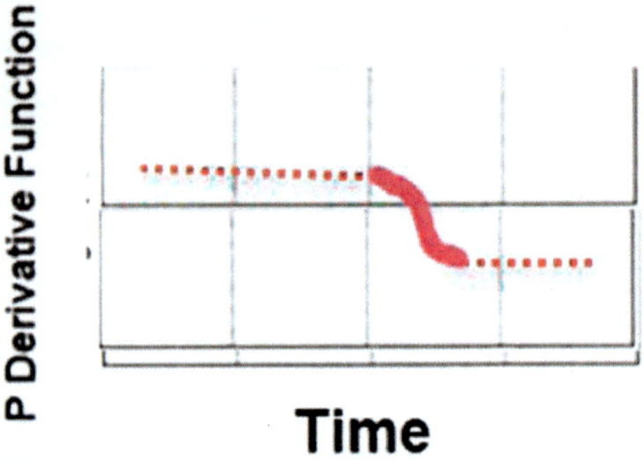

B:

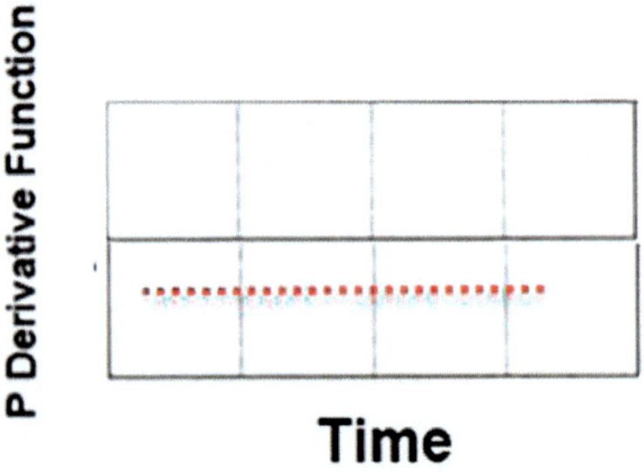

C:

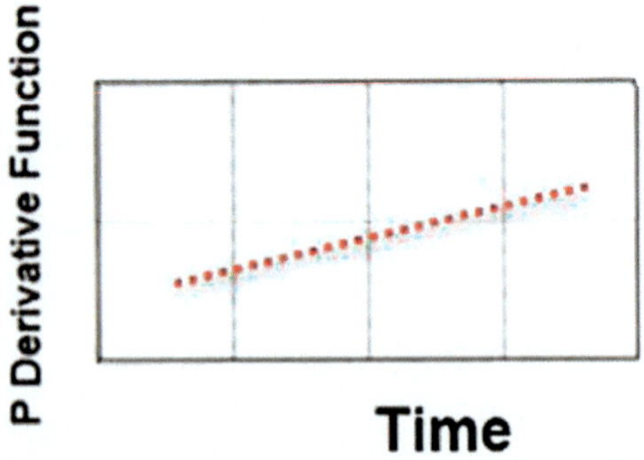

D:

Answer: A is the answer indicating a drop in transmissibility.

Q57-The derivative function plot for a fall-off test is shown below. What is the explanation?

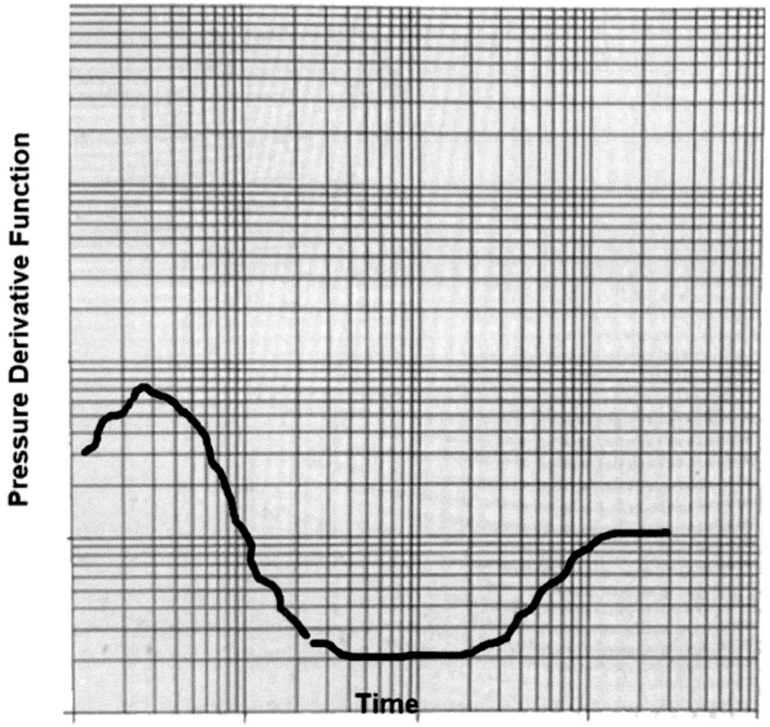

Answer: One possible for this observation is that the well is in a fault block,

Q58-What is the mass flow rate if 10,000 STB/D of water flow into a cubic system in 24 h with a cross section is 20,000 ft^2?

Answer: The ratio of the mass flow rate for a water with a density of 62.4 Lb./cf, over the cross section area is: 5.615 * (10,000 Bbl./D) * (62.4 Lb./Cf)/(20,000) = 175 Lb./D.

Q59-Indicate the well testing concept to the appropriate application

Application	Well Test Method
A-Reservoir Pressure	
B-Skin	
C-Permeability Anisotropy	
D-Sealing nature of a fault	
E-Fluid Bank	

Concept

1-Pulse Test
2-Pressure Buildup
3-Interference Test
4-Injection Well Test
5-Hall plot

Answer: 2, 2, 1, 2, 4

Q60-A well produces at the following rates before it is shut down for a build up test, Indicate a procedure to estimate ko, kw and kg.

qo = Oil rate = 300 STB/D qw = water rate = 200 STB/D qg = gas rate = 120 MSCF/D.

Horner slope = 40 psi/cycle, Rs = Solution GOR=300 SCF/STB, Bo = 1.2, Bg = 0.0001 cf/SCF, Bw = 1

Answer: The general equations is:

$$m = \frac{162.6 \cdot q.\mu.B}{k.h} = \frac{162.6.q}{T} B$$

If we use oil rate and oil properties, we obtain To, the oil transmissibility. If we use water properties, we obtain Tw. For gas, first we need to estimate the free gas flow which can be obtained from:

$$q_g = q_t - q_s$$
$$q_s = R_s \cdot q_0$$

Q61-Given the following curve, estimate the pressure change observed in an observation well 100 ft away from an active well after 10 h of production at 300 Bbl. of oil (no water and no gas production). Estimated transmissibility = 50,000 mD. Ft./cp Estimated Storativity = 0.002 ft/psi rw = 0.5 Bo = 1.2

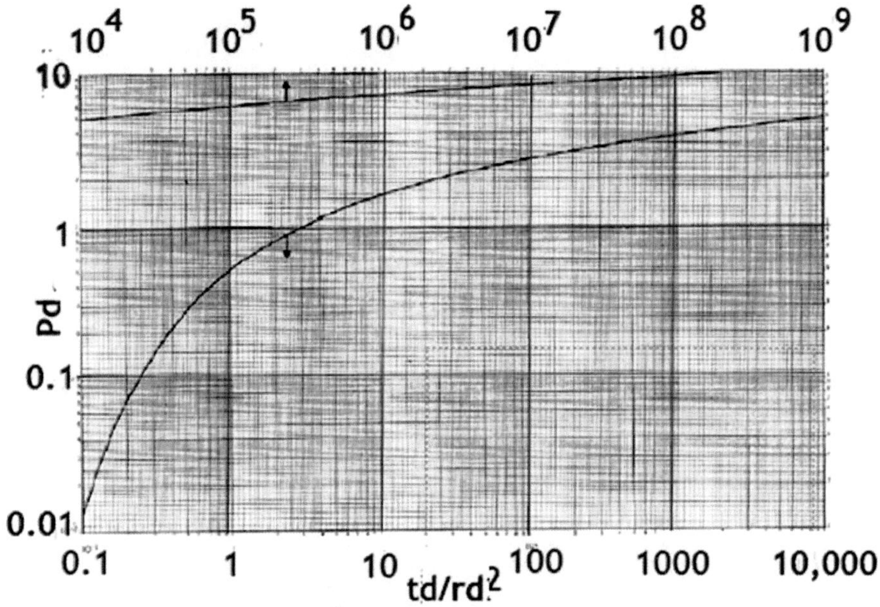

Pd vs. td/rd² (Courtsey SPE Mongragh 5)

Answer: The 10 h is converted to td/rd²

$$td/rd^2 = 0.000264^* T^* t/S^* r2$$

where T = kh/μ and S = φ.ct.h

$$Pd = T \cdot dp/141.2^* q^* B$$

$$td/rd^2 = 0.000264^* 50000^* 10/(0.002^* 10000) = 6.6$$

This corresponds to a Pd = 1.

The pressure change is: 141.2*1.35*300*1.2/50000 = 1.37psi.

Q62-Make a log-log plot showing the changes in well rate vs. time for a hydraulically fractured well in an unconventional reservoir if the reservoir pressure has not changed.

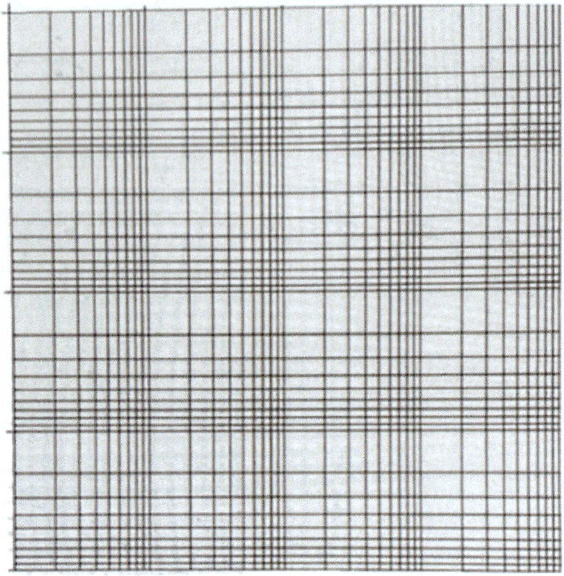

Answer:

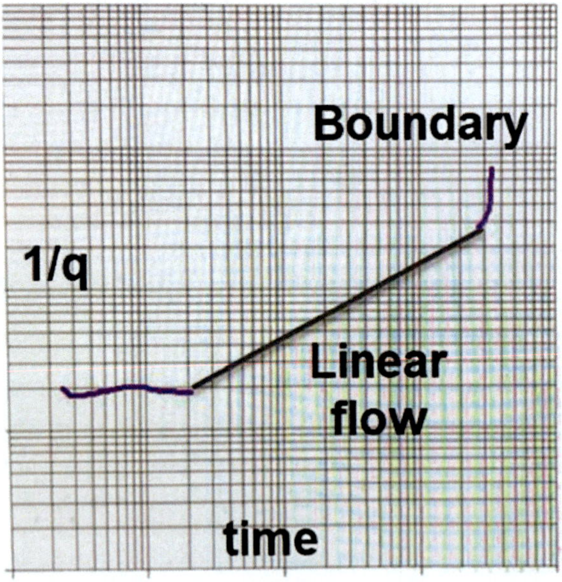

Q63-It takes 48 h of flow to investigate 500 ft. distance from a well. How long would it take to investigate 1000 ft.?

Answer: Under the same reservoir properties, the radius of investigation is proportional to the square root of time. Increasing the distance by factor of 1000/500 = 2 means extending the time by factor of 4.

Q64-Given the following plots, when does the indication of the presence of a fault becomes obvious on the Horner plot? Why?

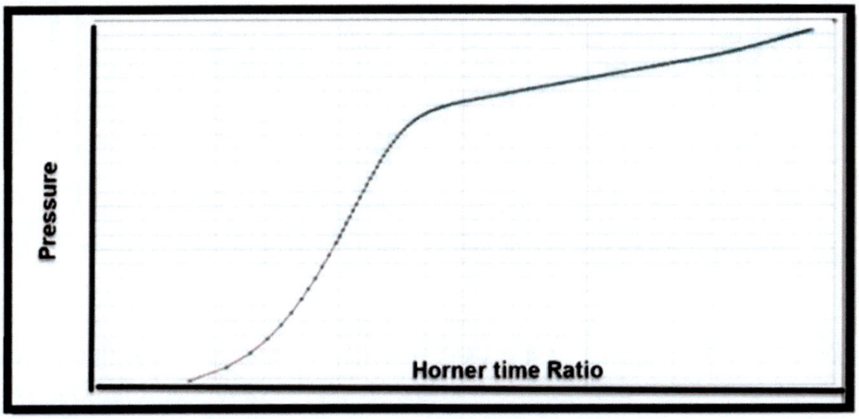

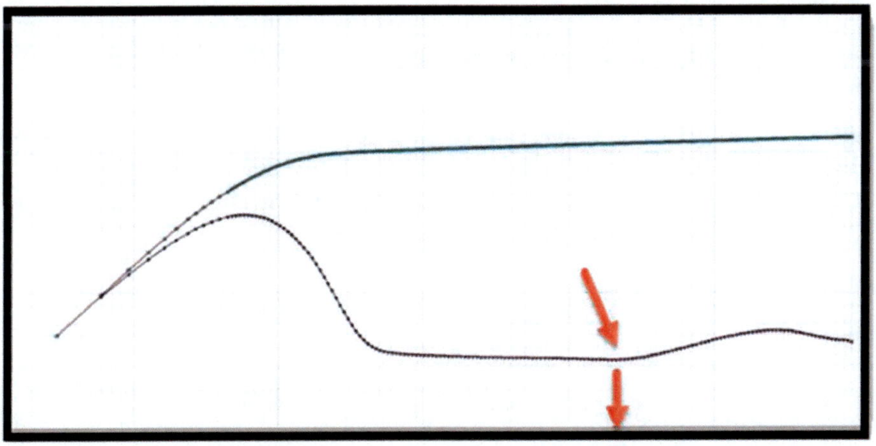

Answer: Any slope changes on the Horner plots associated with a sealing or non-sealing boundary can best be detected on a pressure derivative plot.

Q65-Describe at least two different explanations for the pressure derivative function plot to exhibit the pattern shown below.

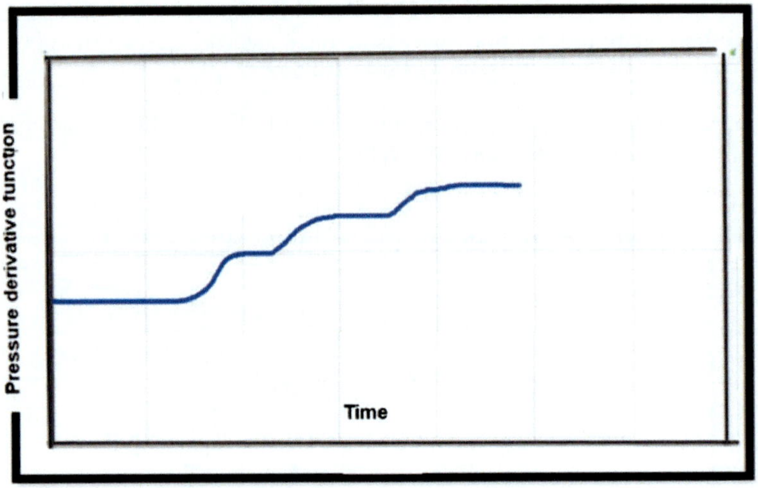

Answer: The slope increase as seen on the pressure derivative function plots indicate several possibilities including either testing a composite system or a layered reservoir.

Q66-If the slope of the Horner plot is 50 psi/ cycle and the skin factor is 20, what is the pressure drop caused by skin?

Answer: Pressure drop caused by the skin factor = 0.87. m.S = 0.87 * 50 * 20 = 870 psi.

Q67-Two wells B and C produce and cause a pressure drop at Well A equally spaced from wells B and C. Well B produces at 500 STB/D and Well C produce at 1000 STB/D. If the total pressure drop is 150 psi, what is the pressure drop caused by well B? Assume an infinite acting homogeneous reservoir.

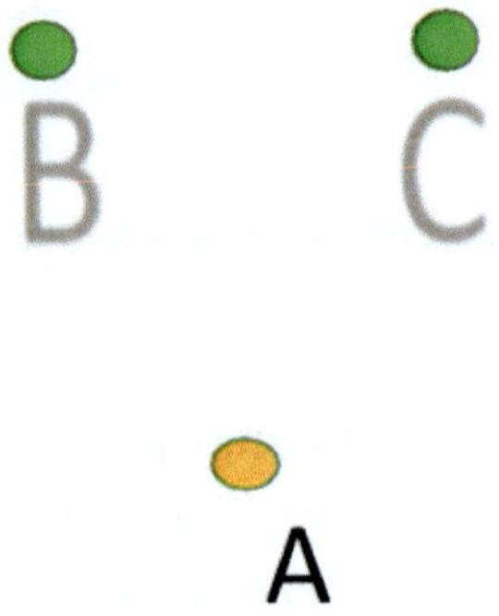

Answer: As pressure drop is directly proportional to q, it will be 100 psi pressure drop caused by well C and 50 psi cause by well B.

Q68-If it takes 300 h of flow at a well to observe a pressure deviation from infinite acting for a circular reservoir where the drainage radius is 700 ft, and the transmissivity is 50,000 md ft./cp, what is the estimated storativity? rw = 0.5 ft.

Answer:

$$t_D = 0.25(re/rw)^2 = 0.25(r/rw)2 = 0.25 * (700/0.5)2 = 490{,}000.$$

$$t_D = 490{,}000 = 0.000264 * \left(\frac{T.t}{S * rw^2}\right)$$

$$S = 0.0323\text{ft./psi}$$

Q69-Well A is produced at 500 STB/D for 24 h. Well B is an observation well. Assume an infinite size reservoir what is the expected pressure drop in well B after 24 h. K = 30 mD., h = 150 ft. $\varphi = 0.2$ rw = 0.5 ft Ct = 30e-6 Viscosity = 1.5 cp B = 1.2 d = 200 ft.

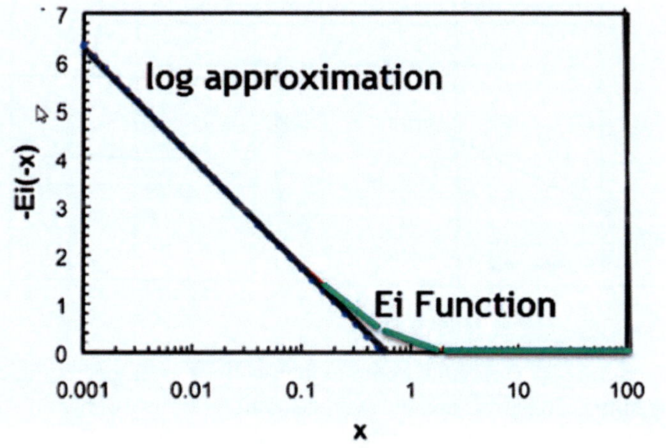

$$\text{Ei}\left(\frac{-948.\varphi.\mu.Ct * r * r)}{kt}\right) = \text{Ei}(-0.474) = -0.6$$

$$\text{dp} = \frac{-70.6 * 500 * 1.5 * 1.2}{30 * 150} * (-0.6) = 6.77\,\text{psi}$$

Q70-Show the expected Horner plot and the pressure derivative plot for a well with no skin and no wellbore storage but with boundary effects.

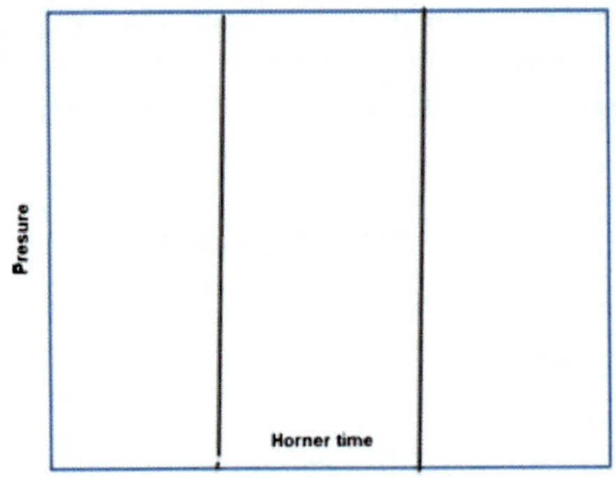

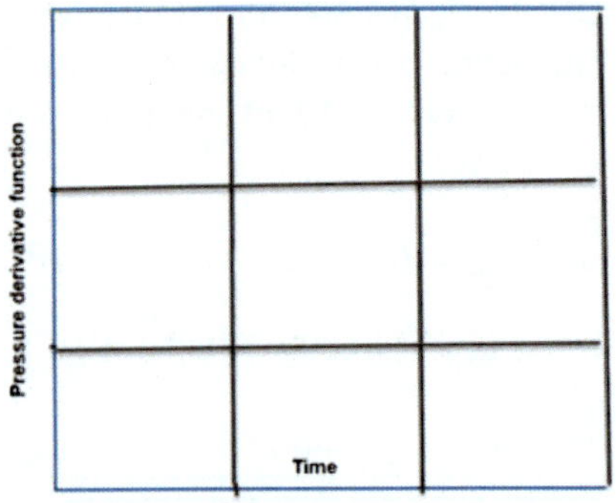

Answer:

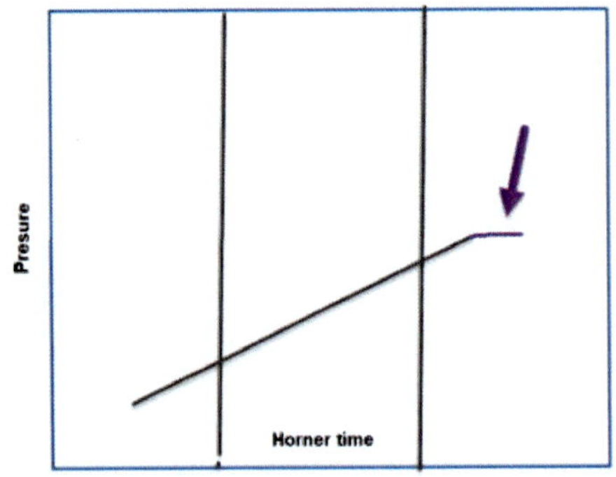

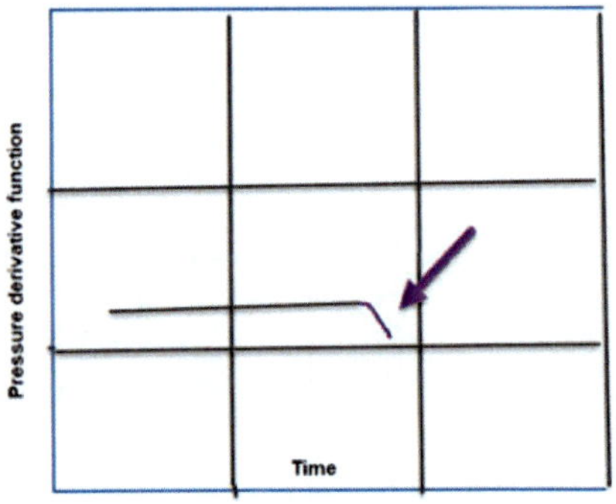

Q71-If the rate is 30 STB/D, compute the wellbore storage coefficient when the tubing capacity is 50 barrels and the average compressibility of the fluid in the tubing is $50E^{-6}$. How does it compare with what is obtained from a test as shown below? $B0 = 1.1$ RB/STB.

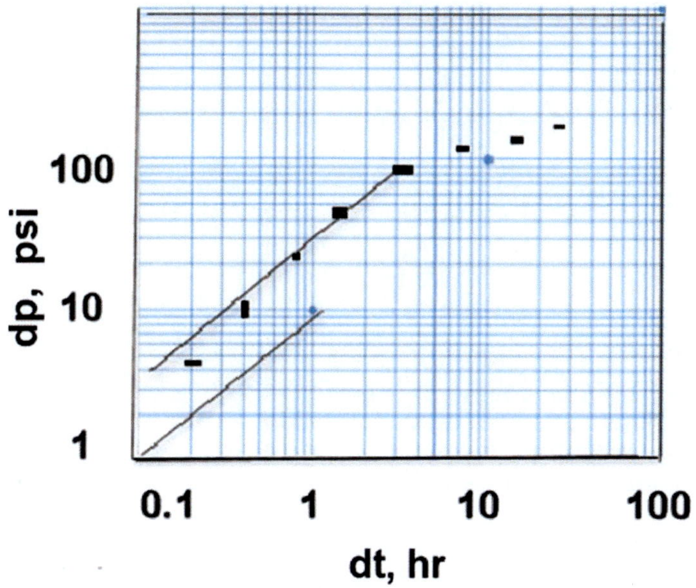

Answer: On the 45° line, we can read any points such as:dp = 30 dt = 1 from the log-log plot Cw = $q.B\,(dt/dp)/24$.

cw = 50 * 50e − 6: Cw = $q \cdot B\left(\frac{1}{24}\right)\left(\frac{1}{30}\right) = 30 * 1.1/720 = 0.0458$RB/psi

From the wellbore data Cw(well) = V*Cw = 50*50e − 6 = 0.0025RB/psi

Higher Cw from well test indicates leaks to outside the tubing and potentially the formation.

Q72-The seismic and a pressure buildup can both see the presence of a nearby fault, but a pressure build up can identify a property of the fault that cannot be obtained from seismic. What is that property?

Answer: Transmissibility

Q73-If we use a silicon-sapphire gauge in the range of 0-10,000 psi, for a reading of 1234. psi, the correct pressure may have been:

A—1239.
B—1237.
C—1244.
D—1243.

Answer: B (for a 10.000 full-scale the accuracy is +/− 3 psi).

Silicon-Sapphire	
Pressure accuracy	0.03% full scale
Pressure Resolution	0.0003%

(continued)

(continued)

Silicon-Sapphire	
Pressure Drift	< 3 psi/year

Q74-The following condition

$$\frac{\partial p_D}{\partial t_D} = \text{constat}$$

Means:
A—Steady State
B—Transient
C—Semi steady State
D—Constraint Pressure boundary

Answer: C

Q75-A well is flowing at 300 STB/D. Assume Skin = 0 and an infinite acting reservoir. If the flow rate increases to 3000 STB/D, the pressure drop at the well is increased by 39 psi. for the same duration What is the total pressure drop?

Answer:

1—$P_{i\_}P_{wf}^{(t)} = \frac{162.6q\mu B}{k.h}\left\{\log\left(\frac{kt}{\varphi\mu crw^2}\right) - 3.23 + 0.87S\right\}$

2—$P_i - P_{wf}^{(t)} = \frac{162.6q\mu B}{k \cdot h}\left\{\log\left(\frac{kt}{\varphi\mu crw^2}\right) - 3.23\right\}$

3—$P_i - P_{wf}^{(t)} = \frac{162.6q\mu B}{k \cdot h}\left\{\log\left(\frac{kt}{\varphi\mu crw^2}\right) - 3.23\right\}$

Divide equation 2 by equation 3. The pressure drop for the 300 STB/D is: Pi-P_{Wf} = 0.39 psi

Q76-Given the following graph, estimate the exact dimensionless time when the solution for a bounded reservoir with re/rw = 2000 deviates from the infinite acting. Show the method used.

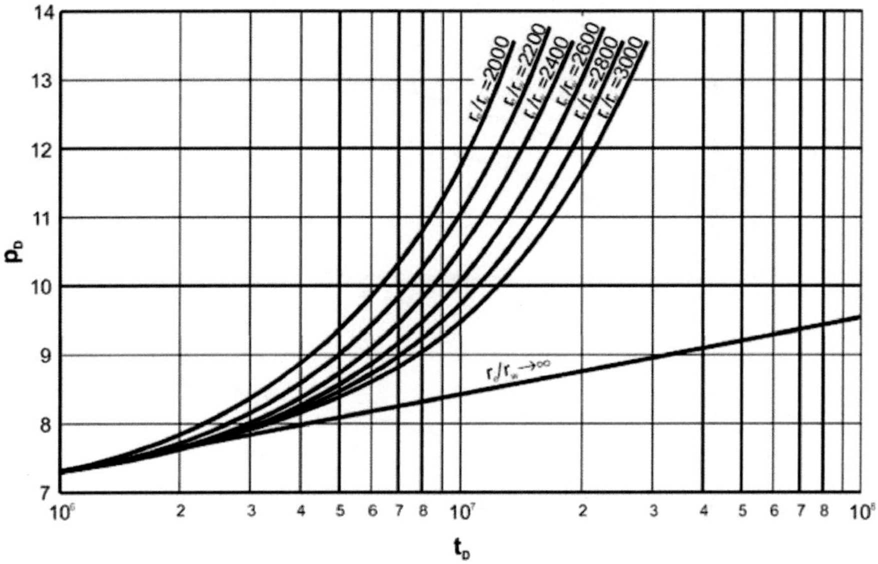

Courtesy SPE Monograph No. 5 by Earlougher
Pressure Response in a Bounded Circular Reservoir

Answer: t_D = 0.25 * (re/rw)2 = 0.25 * (2000)2 = 1,000,000.

Q77-A new gas reservoir has been discovered with the initial pressure of 3464 psia. The volumetric estimate corresponds to a gas in place of 18 BCF. There are 10 wells drilled there and the back pressure equation for all 10 wells is q = 0.0012*(P2-Pwf2) 0.8 (q is in MSCF/D and pressures are in psia).

a—What is the AOF?

Answer: For AOF, Pwf=14.7 psia. q = 0.0012 * (3664 2 – 14.7 2) = 16,109 MSCF/ D

Q78-On the plots shown below for buildup tests indicate the direction of the arrow, increasing or decreasing. S represents the skin factor.

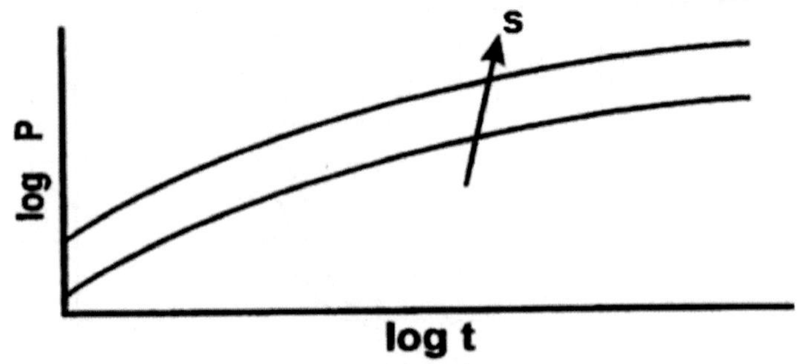

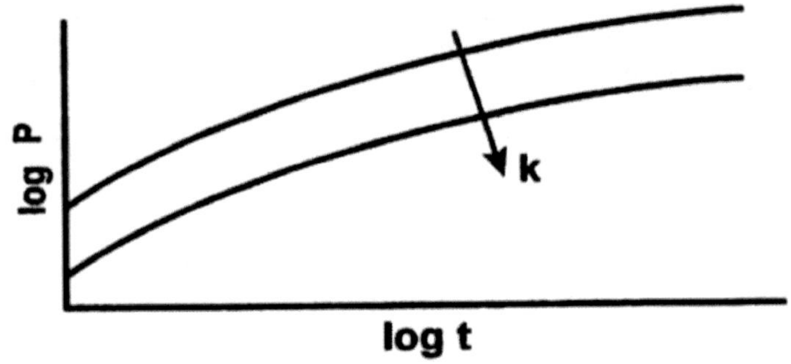

Answer: With higher skin the buildup is slower. As such the direction of arrow is toward lower skin. The lower graph also corresponds to a buildup and for a reservoir with a higher permeability. The buildup is faster and as such the direction of the arrow is toward a lower permeability.

Q79-See the Hall plot shown below from SPE 185694. How do you interpret the plot?

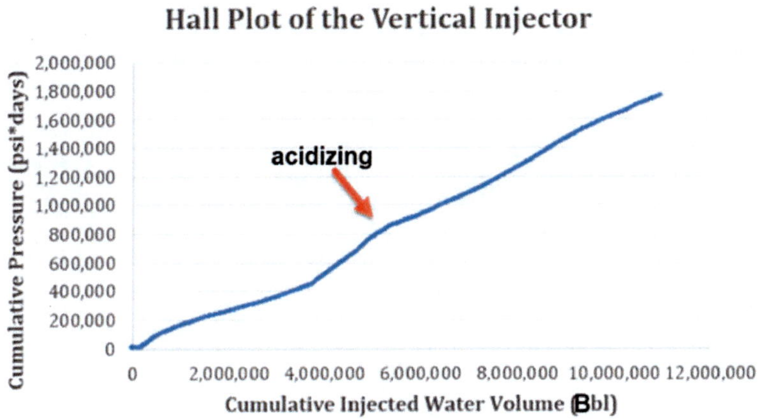

Answer: Slope increases correspond to resistance to injection or skin. Stabilization by acidizing is maintaining injectivity.

Q80-The engineer in charge did not accept this test. Could you identify the reason the test was not acceptable? Mark the portions that may have raised concerns in the mind of the engineer.

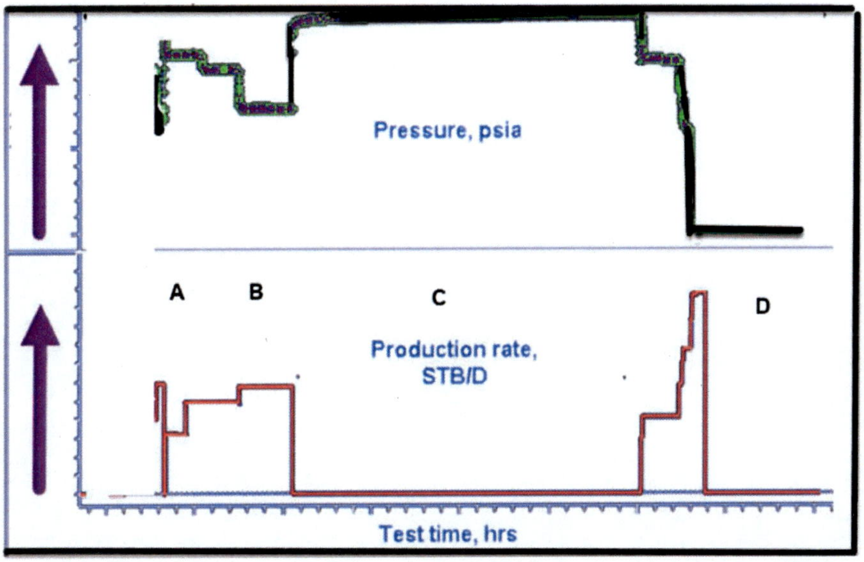

Answer: In section D there is a pressure drop while the rate is zero.

Q81-A Pulse test between two wells was conducted the following data was obtained. Estimate T (transmissibility) and S (Storativity).

Additional data

Distance between the two wells = 424 ft flow rate 100 STB/D B0 = 1.281 rw = 0.5 ft.

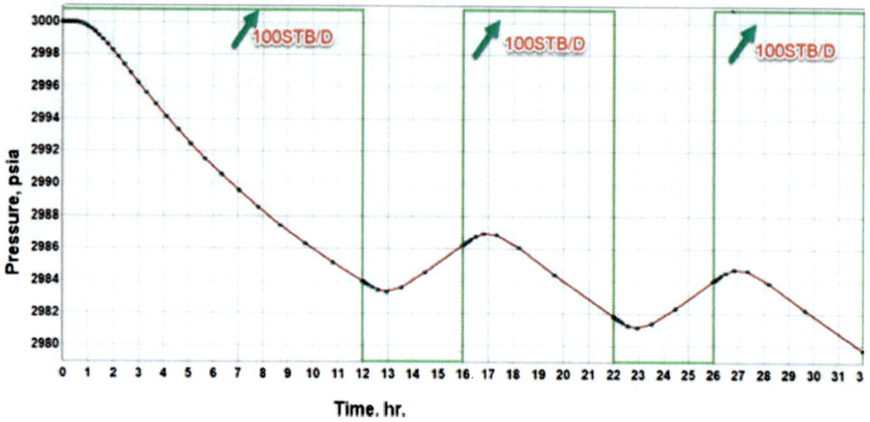

Answer: t_{lag} = 1 hr. and dp = 5 psi. Also, the cycle duration is 10 hours. F' = 4/10 = 0.4 Tld = 1/10 = 0.1

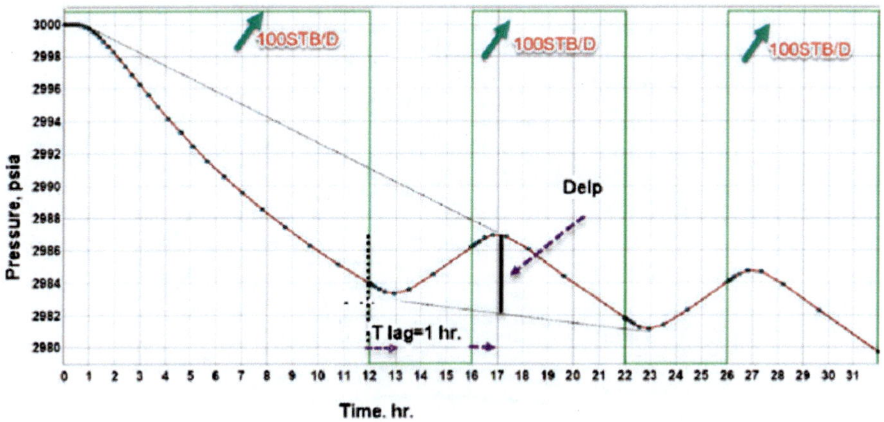

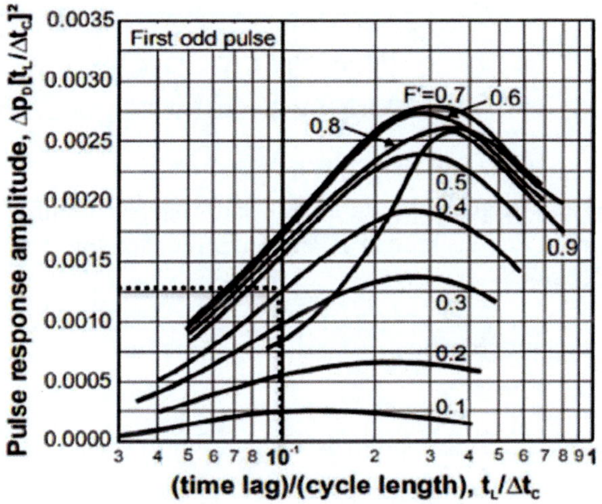

(Courtesy SPE Monograph 5)

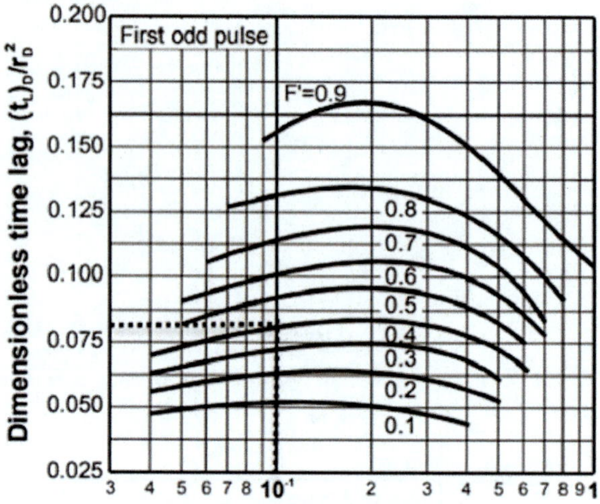

Kamal, M and Brigham, W. E.: Design and Analysis of Pulse Tests with Unequal Pulse and Shut-In Periods Journal of Petroleum Technology 28 (02): 205–212. https://doi.org/10.2118/4889-PA-1975.

Pressure function = 0.00125 Time function = 0.081

dp = 5psi

$$p_d = \frac{0.00125}{0.1} = 0.0125 \qquad t_{ID} = 0.081^*(424/0.5)2 = 69$$

$$T = \frac{141.2 * Pd * q * B}{dp} = 45mD.ft/cp$$

$$S = \frac{0.0000264 * 45 * 1}{69 * 0, 25} = 0.00068$$

Q82-After 80 h of flow in an infinite acting reservoir, the pressure distribution around the producing well is as shown below and the Horner plot generates a straight line with a slope of 50 psi/cycle. What is the skin factor?

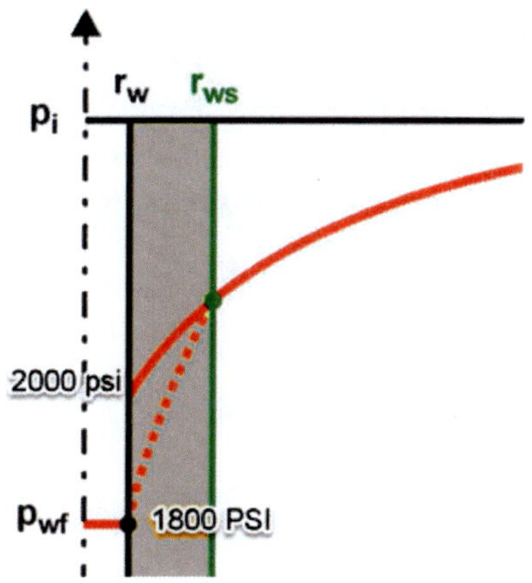

Answer: Pressure drop by skin is (2000–1800) psi. = 0.87.m.S
S = 4.59

Q83-Mark parameter (s) that could not be obtained directly from pressure transient tests:

A—Formation thickness
B—Drainage volume
C—Near wellbore damage
D—Turbulent flow

Answer: A

Q84-What type of the boundary condition results on the three curves being exactly the same in terms of pressure distribution? Assume radial flow, homogeneous reservoir.

A—Infinite Acting
B—Closed Boundary
C—Constant pressure boundary
D—Constant non zero gradient at the boundary

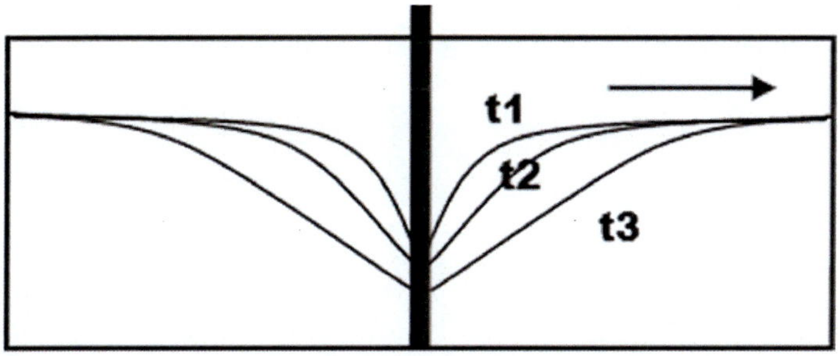

Answer: A infinite acting.

Q85-A new well is placed on production at a constant rate for a period of 10 days. For an oil viscosity of 4 cp this corresponds to a radius of investigation of 150 ft.

How far would have been investigated if the fluid viscosity were 1 cp? Assume radial flow, and homogeneous reservoir.

A—600 ft
B—150 ft
C—300 ft
D—75 f

Answer: C: $r = \sqrt{\frac{c}{\mu}}$ when viscosity (μ) is reduced by a factor of 4 the (r) becomes larger by a factor of 2.

Q86-In in infinite acting reservoir, if a well is on production (q) and pressure drops (dp) are measured at a very large distance, what is the relation between dp versus q?

Assume radial flow, homogeneous reservoir.

A—$dp = a \cdot q + b$.
B—$dp = a \cdot q$.
C—$dp = 1/q$.
D—$dp =$ is independent of q.

Answer: D

Q87-A well is placed on production. Measurement show that after 20 h, a radius of 100 ft. was investigated. How long would it take to investigate 200 ft.?

Assume radial flow, homogeneous reservoir.

A—10 ours
B—Cannot be estimated with the information given
C—40 h
D—80 h

Answer: D $d = \sqrt{c \cdot t}$ whem d increases by a factor of 2, t must increase by a factor of 4.

Q88- From pressure buildup test a wellbore storage coefficient of 0.001 bbl./psi has been estimated. Hanging in the hole is 2 1/16 "tubing (2.063) OD. The effective storage volume is 2 bbl. The effective borehole fluid compressibility is:

A = cannot be estimated

$$B = 50e - 4$$

$$C = 62.5e - 5$$

$$D = e - 5$$

Answer: B

$$Cw = 2 * C \quad C = \text{ compressibility } = 0.005\text{Bbl./psi}$$

Q89-Given the following data, estimate the back pressure isochronal equation and the AOF. Pi = 1952 psig.

Test duration 3 h	
q_{sc} (MSCF/D)	Pwf, psia
2600	1793
3300	1757
5000	1623
6300	1505
Test duration 72 h	
6000	1151

Answer:

Test duration 3 h		
q_{sc} (MSCF/D)	Pwf, psia	$Pr^2 - Pwf^2$
2600	1793	595,455
3300	1757	723,255
5000	1623	1,176,175
6300	1505	1,545,279
Test duration 72 h		
6000	1151	2,485.503

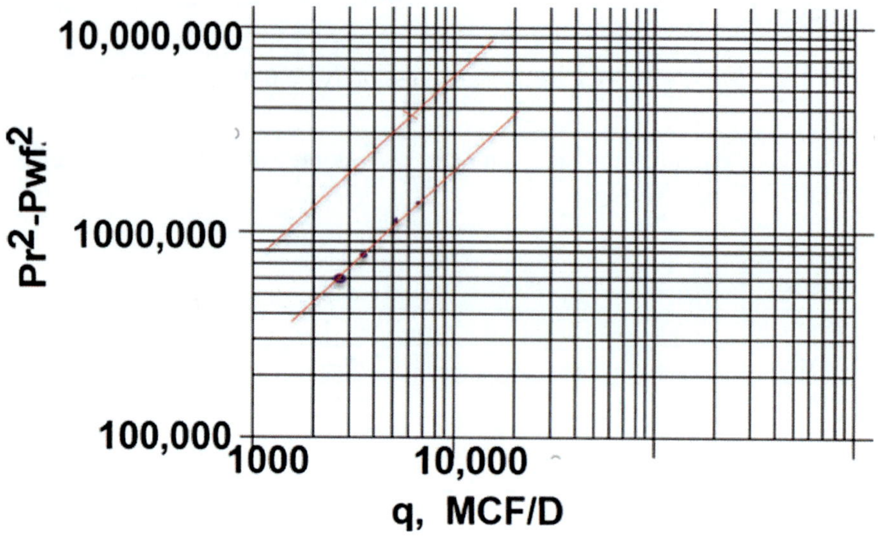

$$n = 0.55 \quad q = C\left(P^2 - p_{wf}^2\right)^{0.55}$$

$$C = 2.48 \quad AOF = 2.48\left(1952^2 - 14.7^2\right)^{0.55}$$

$$\text{AOF} = 10,326 \quad MSF/D$$

Q90-Fluid levels measurements in a rod pump well show after shutting the well for 100 h the level does not change and is in static conditions 240 ft. above the mid point perforation. What is the estimated static reservoir pressure if prior to the shut in the well produced 14 API gravity oil with a water cut of 60%.

Answer: API gravity = ($141.5/\gamma_0 - 131.5$)

For API gravity of 14, the oil specific gravity is $\gamma_0 = 0.972$.

Water pressure gradient is 0.433 psi/ft.

Sp. Gravity of oil = (131.5 + 14)/141.5 = 0.972

Pressure of the fluid column = (0.433 * 0.6 + 0.4 *0.972 *0.433) * 240 = 102 psi

Q91-Using the data DST shown below, write down three important observations.

DST Data
nterval -
4022-4048 ft

IHP=2053	FHH=2036
Flow=20 min	Flow =60 min
IFP=228 psi	
FFP=639	IFP=470 psi
Time closed=44 min	FFP=1062 psi
ISIP=1677	Time closed=91 min
	FSIP=1694

187 ft. Dry gas

248 ft .Gas Cut Mud

Answer:

1—Difference between Initial hydrostatic head and final hydrostatic head

2—Differece between initial flop pressure in the first flow and initial flow pressure for the second test.

3—Difference between initial and final shut-in pressure,

Possibly gas has entered the cushion fluid and also the mud column.

Q92-Consider the type of graph presented by Kamal and Brigham. An engineer is assigned to design a pulse test between two wells. The pulsing well is planned to inject water at 480 reservoir Bbls/d. If the estimated water transmissibility is 1800 mD. ft./cp a. cycle time of 10 h and pulse duration of 5 h, what would the estimated lag time if a pressure pulse of 1 psi is to be expected during the first odd pulse?

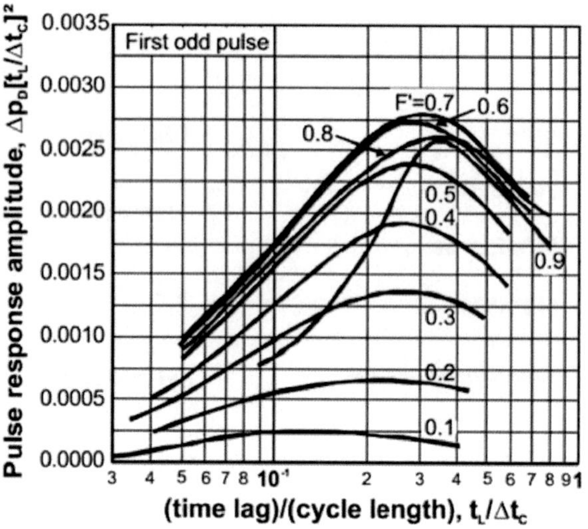

(Courtesy SPE Monograph 5)

Answers:

$$\mathrm{F} = 5/10 = 0.5 \quad \mathrm{Pd} = \frac{T \cdot Dp}{141.2 * q * B} = \frac{1800 * 0.1}{141.2 * 480 * 1.2} = 0.0022$$

$$\frac{tl}{tc} = 0.18 \text{ or } 0.45$$

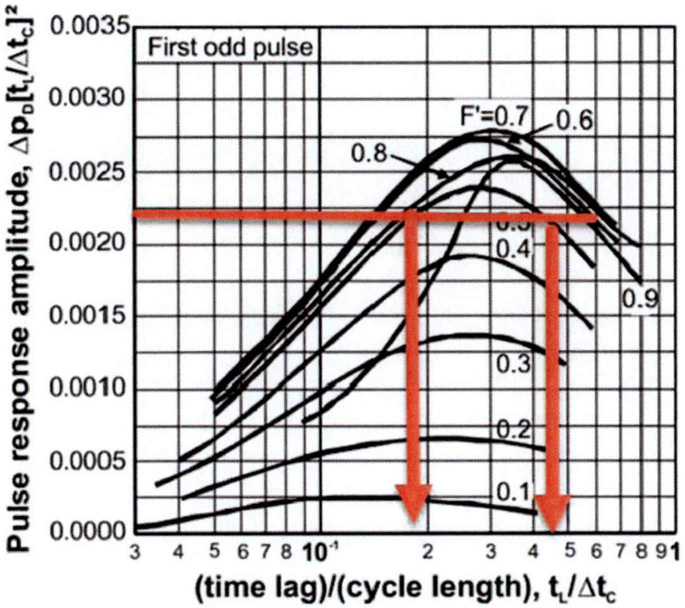

(Courtesy SPE Monograph 5)

Q93-Show derivative plots for a water injection well that after 5 years shows evidence of scale buildup around the well.

Answer:

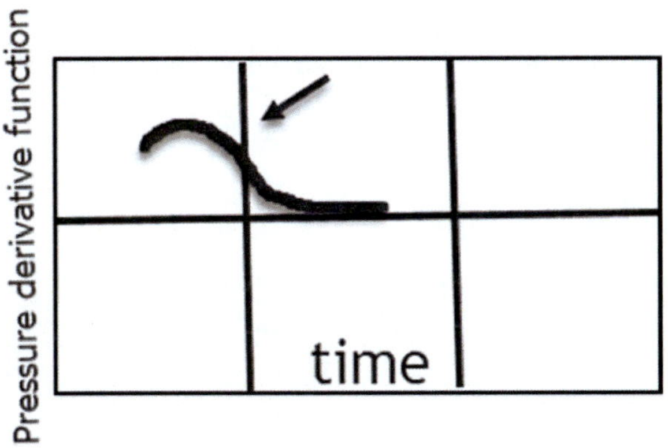

Q94-The results of a drawdown test are matched with the infinite acting model as shown below. The flow rate was 500 res Bbls/D. Estimate T(transmissibility) and the wellbore storage constant. B = 1.2 RB/STB.

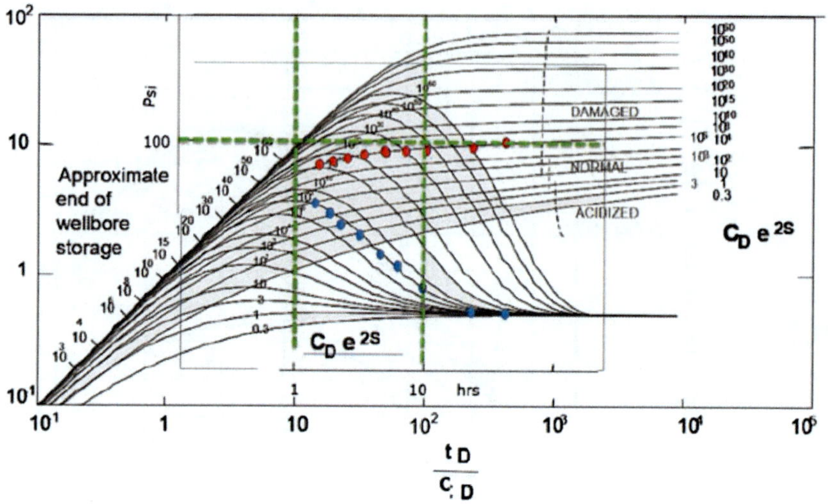

(Type curve courtesy Bourdet et. al 1983 World Oil)

Answer:

Match point : $C_D \cdot e^{2S} = 10^6$

$$\text{Pd} = 10 \quad \text{dp} = 100 \quad p_D = \frac{T dp}{141.2 q B} \quad \text{T} = 8472\text{mD.ft./cp}$$

$$\text{td/Cd} = 100 \quad \text{dt} = 10 \quad \text{td/CD} = 0.000264^* T^* \text{t}/(0.894^*\text{C})$$

$$100 = 0.000264 * 8472 * 10/(0.894^*\text{C}) \quad \text{C} = 0.25\text{Rb/psi}$$

Q95-A hydraulically fractured well is monitored and the plot of dp/q with time is shown below. Explain the plot.

Answer: Linear flow indicative of extensive frac job.

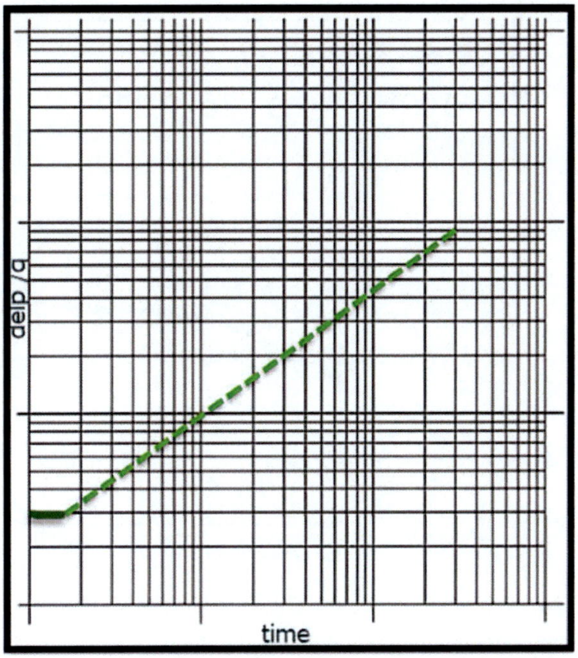

Q96-What does the following Hall plot indicate if the reservoir pressure is assumed constant?

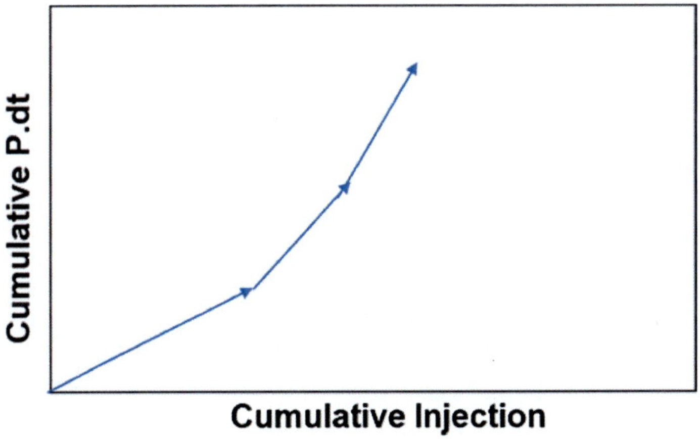

Answers: Drop in injectivity.

Q97-A pressure build up test was conducted in NFR and the following data were obtained. Estimate ω and λ using the *Uldrich- Ershaghi method.

Ct=6e-5 1/psi rw=0.3
h=100ft μ=1.1 Bo=0.372

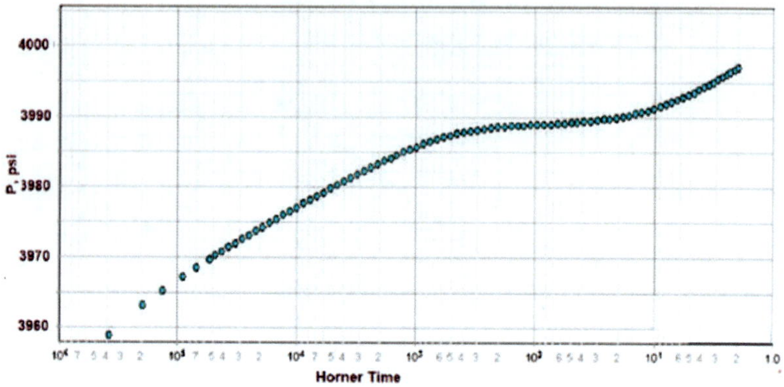

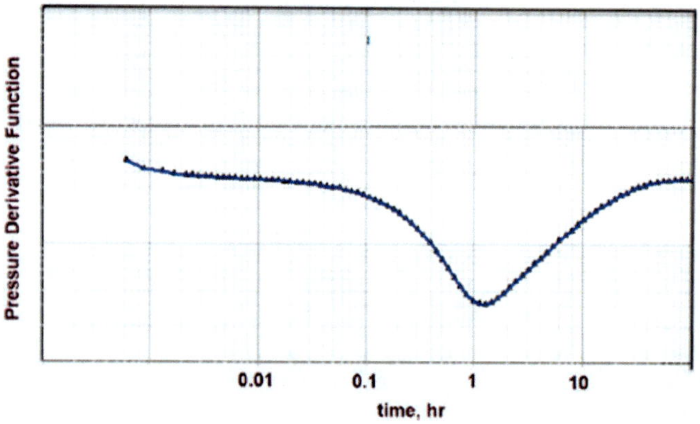

Answer:

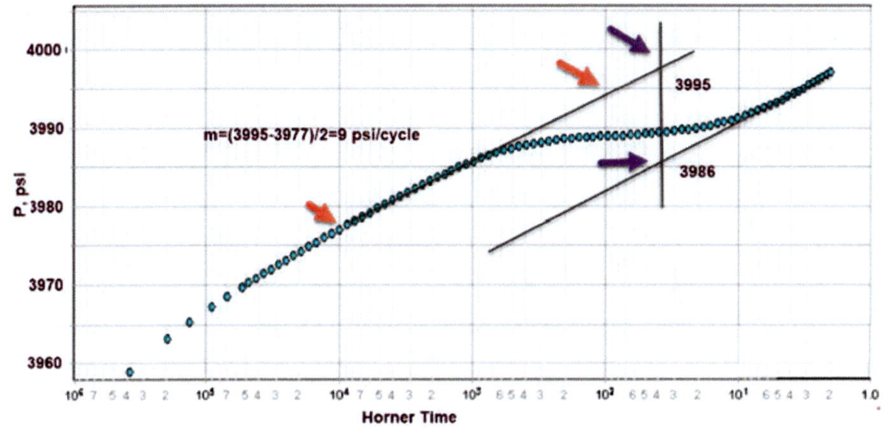

m = 9psi.cycle dp = 9 psi

$$\omega = 0.1 \quad ko = \frac{(162.6).q.\mu.B}{m \cdot h} = 136\,\text{mD}$$

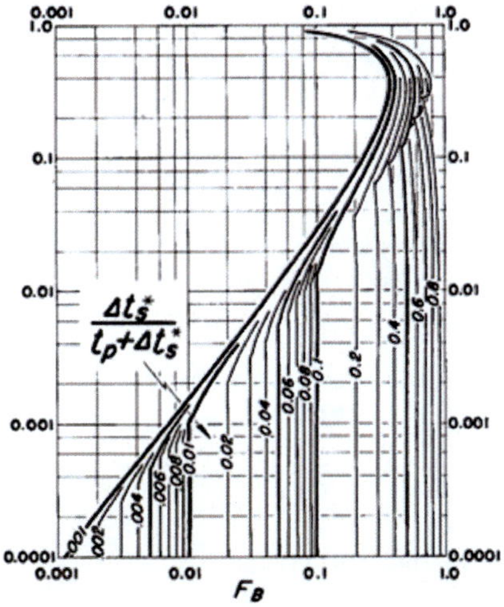

(Courtesy Uldrich, Ershaghi SPE Paper 7142}

FB = 0.3

$$\frac{\lambda}{(\varphi ct)f + (\varphi ct)ma} = \frac{F_{B^{\mu}rw^2}}{0.000264k\Delta ts^*}$$

$$\lambda = \frac{0.00006 * 0.3 * 1.1 * (0.3)(0.3)}{0.000264 * 136 * 1.5} = 0.000033$$

*Uldrich, D.O., Ershaghi, I.: A Method for Estimating the Interporosity Flow Parameter in Naturally Fractured Reservoir. SPE-AIME (1979)

Q98-We plan to run a pressure buildup test. How long should we flow the well and what should be the duration of the build up test to extend the data beyond that affected by wellbore storage and skin.

$$\text{Estimatedwellborestorage} = 0.2\text{bbl./psi} \quad \text{s} = 30$$

$$\text{Estimatedtransmissivity} - 40{,}000\text{md.ft/cp}$$

Flow time:

$$t = \frac{C * (200{,}000 + 120{,}000 * S)}{\text{T}} = 2.8\text{ h}$$

Shut in time: t should be larger than $dt = \frac{(647{,}000+37{,}700*\text{S})*\text{C}}{T} = 8.9\,\text{h}$.

Q99-Consider the following well location. If the well is flowed for 20 h, which boundary will be investigated on the test data. T = 20,000 mD. Ft/cp Storativity = 0.001 ft/psi rw = 0.5 ft

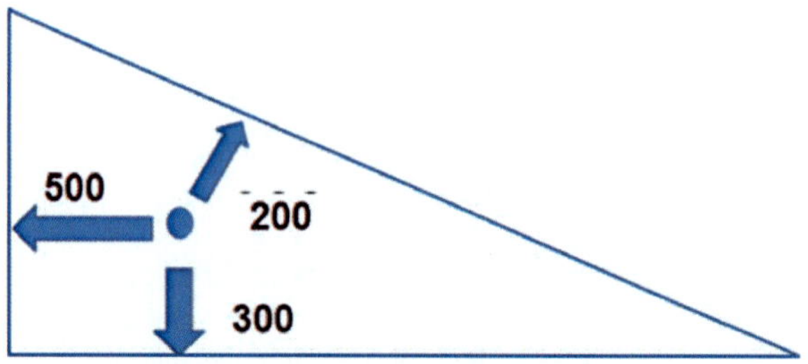

Answer: The closest boundary, 200 ft.

Q100-Examine the following DST charts. Which charts show little evidence of flow?

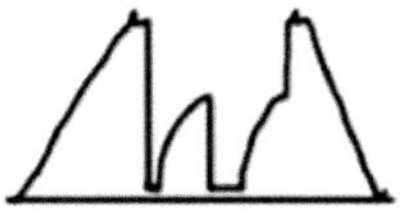

A

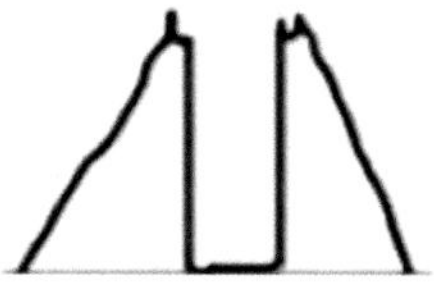

B

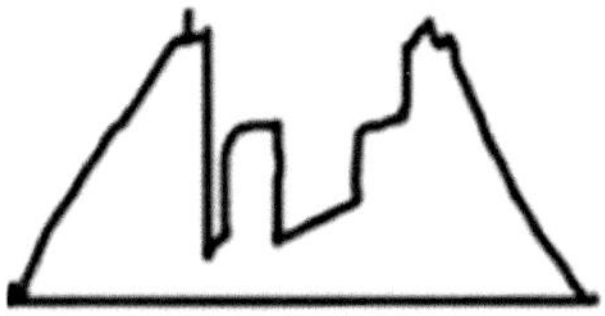

C

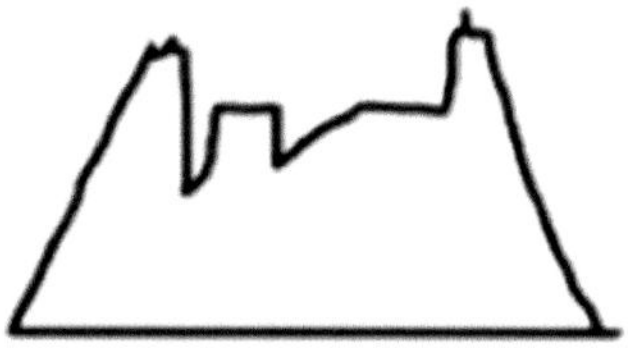

D

Answer: A and B.